Novel Trends in Brain Science

Minoru Onozuka, Chen-Tung Yen
Editors

Novel Trends in Brain Science

Brain Imaging, Learning and Memory, Stress and Fear, and Pain

Minoru Onozuka
Professor
Department of Physiology
  and Neuroscience
Kanagawa Dental Collge
82 Inaoka-cho, Yokosuka
Kanagawa 238-8580, Japan
onozuka@kdcnet.ac.jp

Chen-Tung Yen
Professor
Institute of Zoology
National Taiwan University
No. 1, Section 4
Roosevelt Road
Taipei 106
Taiwan ROC
ctyen@ntu.edu.tw

Library of Congress Control Number: 2007934530

ISBN 978-4-431-73241-9 Springer Tokyo Berlin Heidelberg New York

This work is subject to copyright. All rights are reserved, whether the whole or part of the material is concerned, specifically the rights of translation, reprinting, reuse of illustrations, recitation, broadcasting, reproduction on microfilms or in other ways, and storage in data banks.
The use of registered names, trademarks, etc. in this publication does not imply, even in the absence of a specific statement, that such names are exempt from the relevant protective laws and regulations and therefore free for general use.
Product liability: The publisher can give no guarantee for information about drug dosage and application thereof contained in this book. In every individual case the respective user must check its accuracy by consulting other pharmaceutical literature.

Springer is a part of Springer Science+Business Media
springer.com
© Springer 2008
Printed in Japan

Typesetting: SNP Best-set Typesetter Ltd., Hong Kong
Printing and binding: Kato Bunmeisha, Japan

Printed on acid-free paper

Preface

An enormous number of neurons in the brain form neural circuits by creating synaptic bonds with other neurons. Many of those circuits come together and function as a system. To clarify the function of the human brain, a study needs to be developed as a whole, not merely at the molecular, cellular, or systemic level.

With the development of neural science, knowledge of the molecules and neurons that comprise the brain has increased exponentially in the last two decades. The mysteries of human brain function are being solved in sequence with the use of the latest medical technologies such as PET, fMRI, MEG, and TEM, and in this way, research in neural science is making rapid progress.

Springer Japan has given me the opportunity to publish this book, *Novel Trends in Brain Science*, which introduces the most important and latest knowledge in neural science on brain function imaging, learning, memory, emotions, and pain. In particular, the relationship between well-developed oral cavities and highly advanced brain function is an extremely new and unique area of study and is expected to become an entirely new field.

As of 2006, two Japanese–Taiwanese joint conferences on neural science have been held, in Gifu Prefecture (Japan) last year and in Taipei this year. In this book I have collected original research introduced by neuroscientists representing both countries at those conferences, and I am confident that the work brought together here will promote further research on the human brain.

We express our hearty thanks to staff members at Springer for their efforts in publication, and give special thanks to Nobuyuki Harikae and Ryota Sasaki at LOTTE Co., Ltd. for their support in the editing of this book.

Minoru Onozuka and Chen-Tung Yen

Contents

Preface V

List of Contributors IX

Part I Main Focus
Section I Brain Imaging

1 Diffusion Magnetic Resonance Imaging in Neuroimaging
  W.-Y. I. Tseng and L.-W. Kuo 5

2 Overview of MR Diffusion Tensor Imaging and Spatially
  Normalized FDG-PET for Diffuse Axonal Injury Patients with
  Cognitive Impairments
  A. Okumura, J. Shinoda, and J. Yamada 25

3 Transcranial Magnetic Stimulation in Cognitive
  Brain Research
  S-H. A. Chen 37

4 Spectral Analysis of fMRI Signal and Noise
  C.-C. Chen and C.W. Tyler 63

5 Magnetoencephalography: Basic Theory and Estimation
  Techniques of Working Brain Activity
  Y. Ono and A. Ishiyama 77

Part II Related Topics
Section I Learning and Memory

6 Interactions Between Chewing and Brain Activity in Humans
  M. Onozuka, Y. Hirano, A. Tachibana, W. Kim, Y. Ono,
  K. Sasaguri, K. Kubo, M. Niwa, K. Kanematsu, and
  K. Watanabe 99

7 **Involvement of Dysfunctional Mastication in Cognitive System Deficits in the Mouse**
K. Watanabe, K. Kubo, H. Nakamura, A. Tachibana, W. Kim,
Y. Ono, K. Sasaguri, and M. Onozuka 115

8 **Cellular and Molecular Aspects of Short-Term and Long-Term Memory from Molluscan Systems**
M. Sakakibara 131

9 **Role of the Noradrenergic System in Synaptic Plasticity in the Hippocampus**
M.-Y. Min, H.-W. Yang, and Y.-W. Lin 149

Section II Stress and Fear

10 **Involvement of the Amygdala in Two Different Forms of the Inhibitory Avoidance Task**
K.-C. Liang, C.-T. Yen, C.-H. Chang, and C.-C. Chen 167

11 **Bruxism and Stress Relief**
S. Sato, K. Sasaguri, T. Ootsuka, J. Saruta, S. Miyake,
M. Okamura, C. Sato, N. Hori, K. Kimoto, K. Tsukinoki,
K. Watanabe, and M. Onozuka 183

Section III Pain

12 **Muscular Pain Mechanisms: Brief Review with Special Consideration of Delayed-Onset Muscle Soreness**
K. Mizumura.. 203

13 **ASIC3 and Muscle Pain**
C.-C. Chen 225

14 **Tail Region of the Primary Somatosensory Cortex and Its Relation to Pain Function**
C.-T. Yen and R.-S. Chen 233

Key Word Index 253

List of Contributors

Chang, C.-H., 167
Chen, Chien-Chung, 63
Chen, Chih-Cheng, 225
Chen, Chun-Chun, 167
Chen, R.-S., 233
Chen, S-H.A., 37

Hirano, Y., 99
Hori, N., 183

Ishiyama, A., 77

Kanematsu, K., 99
Kim, W., 99, 115
Kimoto, K., 183
Kubo, K., 99, 115
Kuo, L.-W., 5

Liang, K.-C., 167
Lin, Y.-W., 149

Min, M.-Y., 149
Miyake, S., 183
Mizumura, K., 203

Nakamura, H., 115
Niwa, M., 99

Okamura, M., 183
Okumura, A., 25
Ono, Y., 77, 99, 115
Onozuka, M., 99, 115, 183
Ootsuka, T., 183

Sakakibara, M., 131
Saruta, J., 183
Sasaguri, K., 99, 115, 183
Sato, C., 183
Sato, S., 183
Shinoda, J., 25

Tachibana, A., 99, 115
Tseng, W.-Y. I., 5
Tsukinoki, K., 183
Tyler, C.W., 63

Watanabe, K., 99, 115, 183

Yamada, J., 25
Yang, H.-W., 149
Yen, C.-T., 167, 233

Part I
Main Focus

Part 1
Main Focus

Section I
Brain Imaging

Diffusion Magnetic Resonance Imaging in Neuroimaging

Wen-Yih Isaac Tseng[1] and Li-Wei Kuo[2]

Summary

Diffusion is one of the basic phenomena in nature and can be quantified with magnetic resonance imaging (MRI) in terms of signal attenuation. Diffusion MRI portrays the geometry of microstructures based on the fact that molecular diffusion varies in different directions and that the geometry can be faithfully reflected by the probability density function (PDF) of the molecular diffusion. The theory of MRI shows that the PDF can be obtained by Fourier transform of the MRI signal in the acquisition space (q-space). A technique that stemmed from this theory, called diffusion spectrum imaging (DSI), is capable of resolving complex axonal fiber geometry at each location in the brain, which consequently affords more accurate results of tractography and white matter integrity. Implementation of DSI, however, requires stringent hardware performance and long scan times, thus hampering its clinical application. To overcome these problems, many novel approaches have been proposed using fewer sampling numbers or lower gradient strengths. The advances in diffusion MRI open a door to the investigation of functional connectivity and white matter abnormalities in neuropsychiatric disorders. A future direction for diffusion MRI is to make this advanced technique more clinically feasible and to discover its clinical values in the diagnosis or treatment of brain disease.

Key words Diffusion, Magnetic resonance imaging, Brain, Neuroscience, Tractography

Introduction

The current view of how the brain works is similar to how a concert is performed. A specific brain function is the result of coordinated actions of different neurons at different locations. To understand the function of the brain, it is necessary to record

---

[1]Center for Optoelectronic Biomedicine and Department of Radiology, National Taiwan University College of Medicine, 1 Jen-Ai Road, Section 1, Taipei 100, Taiwan

[2]Interdisciplinary MRI/MRS Laboratory, Department of Electrical Engineering, National Taiwan University, 1 Jen-Ai Road, Section 1, Taipei 100, Taiwan

the activities over the whole brain in both a spatially and temporally resolved manner. To date, many imaging modalities have been used to investigate brain function, including electroencephalography (EEG), magnetoencephalography (MEG), positron emission tomography (PET), and magnetic resonance imaging (MRI), with their respective advantages and disadvantages [1].

Compared with other imaging modalities, MRI provides mapping of brain function and structure with plausible quality. Because MRI produces no ionizing radiation, it is suitable for experiments in healthy populations or repetitive studies in patients. Since the 1990s, functional MRI (fMRI) has become a major tool for investigating the cognitive function of the human brain [2]. Although fMRI can map brain activation in response to a specific function, it cannot provide information about how different regions of activation are connected. Recently, the advances made in diffusion MRI have allowed us to visualize neural fiber tracts and measure their integrity [3].

Given the notion that neural fiber tracts are the structural basis of functional connectivity, the use of diffusion MRI, probably in combination with fMRI, may shed new light on how the brain works. This chapter familiarizes readers with this emerging technology by discussing the principles of diffusion MRI, the evolution of the technology, its clinical applications, and future directions.

Diffusion Phenomenon

The wiggling motion of pollen in water results from collisions of the surrounding water molecules, which are in constant agitation due to thermal brownian motion. An example of diffusion is a drop of ink in water. The drop gradually expands its territory in the water as time passes. This phenomenon can be described by Fick's law, which relates the flux of ink particles J to the spatial gradient of the ink density ρ

$$\mathbf{J} = -D\nabla\rho \qquad [1]$$

where D is the diffusion coefficient. Due to the law of mass conservation

$$\partial\rho/\partial t = -\nabla \cdot \mathbf{J} \qquad [2]$$

Equation 1 can be rewritten more explicitly as the rate of change in ink density, $\partial\rho/\partial t$, which is proportional to the Laplacian of the density $\nabla^2\rho$.

$$\partial\rho/\partial t = -\nabla \cdot D\nabla\rho = -D\partial^2\rho/\partial^2\mathbf{x} = -D\nabla^2\rho \qquad [3]$$

Setting the density function at time 0 be a Dirac's delta function, $\rho(\mathbf{x}, 0) = \delta(\mathbf{x} - \mathbf{x}_0)$, the solution for Eq. 3 is a Gaussian function of time and space with the width of $(2Dt)^{1/2}$.

$$\rho(\mathbf{x}, t) = \text{Exp}[-(\mathbf{x} - \mathbf{x}_0)^2/(4Dt)]/(4\pi Dt)^{3/2} \qquad [4]$$

From the above derivation we know that the solution to the density distribution function of a species of particles is a Gaussian function. As expected, the Gaussian starts with a spike at $\mathbf{x}_0$ and broadens with time (Fig. 1). This simplifies our measurement of diffusion in terms of the diffusion coefficient because this index suffices to represent the diffusion of a species of particles. Although Eq. 4 describes how the density distribution function of an ensemble of particles evolves over time by diffusion, the same solution is also valid for the behavior of a single particle, which can be described in terms of the self probability density function (PDF).

$$P_s(\mathbf{x}, t) = P_s(\mathbf{x}_0|\mathbf{x}, t) \qquad [5]$$

The meaning of the self PDF is as follows. Given a molecule at position $\mathbf{x}_0$ at time $t = 0$, self PDF is the probability of finding the same molecule at position $\mathbf{x}$ at time t. Therefore, $P_s(\mathbf{x}_0|\mathbf{x}, t)$ can be expressed as:

$$P_s(\mathbf{x}_0|\mathbf{x}, t) = \text{Exp}[-(\mathbf{x} - \mathbf{x}_0)^2/(4Dt)]/(4\pi Dt)^{3/2} \qquad [6]$$

In isotropic diffusion, $P_s(\mathbf{x}_0|\mathbf{x}, t)$ is independent of $\mathbf{x}_0$, and $(\mathbf{x} - \mathbf{x}_0)$ can be replaced by $\Delta\mathbf{x}$. Thus, Eq. 6 can be simplified to:

$$P_s(\Delta\mathbf{x}, t) = \text{Exp}[-\Delta\mathbf{x}^2/(4Dt)]/(4\pi Dt)^{3/2} \qquad [7]$$

Equation 7 allows us to obtain a working definition of the diffusion coefficient by quantifying the "Einstein's length" or "mean square length," $\langle \Delta\mathbf{x}^2 \rangle$, of a particle.

$$\langle \Delta\mathbf{x}^2 \rangle = \int_{-\infty}^{\infty} \Delta\mathbf{x}^2 P_s(\Delta\mathbf{x}, t) d\Delta\mathbf{x} = 6Dt \qquad [8]$$

The Einstein's length in Eq. 8 is in the three-dimensional space; it is 2Dt in one dimension. The diffusion coefficient D can then be determined if we know the Einstein's length of a particle during t.

$$D = \langle \Delta\mathbf{x}^2 \rangle / 6t \qquad [9]$$

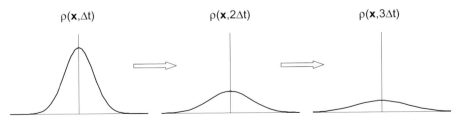

Fig. 1. Change in the density distribution function over time under the diffusion process. It can be described as a Gaussian function that gradually broadens with time

There are three ways to measure the diffusion phenomenon in different diffusion lengths: neutron scattering, MRI, and radioactive tracers, covering nanometer, micrometer, and millimeter ranges, respectively [4, 5]. MRI is the most suitable tool for biomedical application because it has no ionizing radiation and its diffusion sensitivity matches the typical cell size of a living organism [6].

Diffusion Measurement Using MRI

Magnetic resonance imaging measures self-diffusion of water molecules based on the relation between the signal of an ensemble of protons and the translational displacement of water molecules. Hahn first noticed that the signal intensity of a spin echo can be attenuated by the diffusion in the presence of magnetic gradients [7]. It can be understood by the diagram in Fig. 2. In an ordinary spin echo sequence, two magnetic gradients are added: one between the 90° and 180° radiofrequency (RF) excitations, and the other between the 180° RF pulse and spin echo. Because of linearity between the Larmor frequency and the magnetic field, the protons at

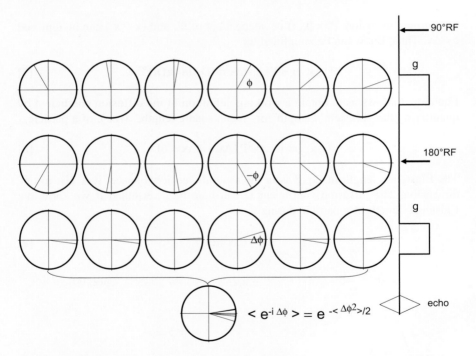

Fig. 2. Mechanism of magnetic resonance imaging (MRI) signal attenuation due to diffusion. The diffusion gradient causes linear phase shifts of transverse magnetization. These phase shifts are refocused by the 180° radiofrequency (*RF*) pulse if there is no diffusion. The diffusion produces phase dispersion of the magnetization (i.e., $\langle e^{-i\Delta\phi} \rangle$, resulting in signal attenuation ($e^{-\langle \Delta\phi^2 \rangle}$). *g*, diffusion gradient

different locations are tagged with different phases by the first gradient and tagged again with additional phases by the second gradient. If water molecules did not undergo diffusion, the molecules stay at the same spots at the times of the two gradients and obtain the same amount of phases with opposing signs due to the reversing effect of the 180° RF pulse. The phases are cancelled, leading to complete refocusing of the spin echo signal. In the presence of diffusion, translational displacement of molecular random walk makes the molecules located at different spots at the times of the two gradients be endowed with different phases. This causes dispersion of the proton spins at the echo time, thus attenuating the signal.

Torrey [8] first introduced the diffusion terms $\nabla \cdot \mathbf{D}\nabla \mathbf{M}$ into the Bloch equation, which governs the diffusive transport of proton spins $\mathbf{M}$.

$$\partial \mathbf{M}/\partial t = \gamma \mathbf{M} \times \mathbf{B} - \begin{pmatrix} \frac{1}{T_2} & 0 & 0 \\ 0 & \frac{1}{T_2} & 0 \\ 0 & 0 & \frac{1}{T_1} \end{pmatrix} \mathbf{M} + \mathbf{M}_0 \begin{pmatrix} 0 \\ 0 \\ \frac{1}{T_1} \end{pmatrix} + \nabla \cdot \mathbf{D}\nabla \mathbf{M} \qquad [10]$$

where $\mathbf{B}$ is the magnetic field experience by the protons $\mathbf{B} = \{0, 0, B_0 + \mathbf{x} \cdot \mathbf{g}(t)\}$, and γ is the gyromagnetic ratio of proton spins [8]. With some derivation, one can obtain the following result

$$\partial \mathbf{M}/\partial t = -4\pi^2 \mathbf{M}(t)[\mathbf{q}^T(t) \cdot \mathbf{D} \cdot \mathbf{q}(t)] \qquad [11]$$

where $\mathbf{q}(t) = \bar{\gamma} \int \mathbf{g}(\tau)d\tau$ is the diffusion sensitivity, which is the spatial modulation of the transverse magnetization created by the diffusion gradient $\mathbf{g}$. Here the unit of $\mathbf{q}$ is length inverse, hence $\bar{\gamma} = \gamma/(2\pi)$. The solution to Eq. 11 is:

$$\mathbf{M}(t) = \mathbf{M}(0)\mathrm{Exp}\left[-4\pi^2 \int_0^t \mathbf{q}^T(\tau) \cdot \mathbf{D} \cdot \mathbf{q}(\tau)d\tau\right] \qquad [12]$$

Equation 12 states that the attenuated signal M(t) is related to the diffusion coefficient $\mathbf{D}$ and diffusion sensitivity $\mathbf{q}$ in a nonlinear (i.e., exponential decay) fashion. Note also that $\mathbf{D}$ in Eq. 12 is not a single scalar. To account for the restrictive diffusion, which may be different along different directions, a more general form, namely a rank two tensor, is used. We can rewrite the integrand in terms of 1×9 component matrices $\mathbf{b}$ and $\mathbf{d}$

$$\ln[\mathbf{M}(t)/\mathbf{M}_0(t)] = -4\pi^2 \int_0^t \mathbf{q}^T(\tau) \cdot \mathbf{D} \cdot \mathbf{q}(\tau)d\tau \qquad [13]$$

$$= -4\pi^2 \left[\int_0^t \mathbf{q}(\tau) \otimes \mathbf{q}^T(\tau)d\tau\right] \cdot \mathbf{D} \qquad [14]$$

$$= -\mathbf{b} \cdot \mathbf{d} \qquad [15]$$

where $\mathbf{b}^T = 4\pi^2 \int_0^t (q_x q_x, q_x q_y, q_x q_z, q_x q_y, q_y q_y, q_y q_z, q_x q_z, q_y q_z, q_z q_z) d\tau$ and $\mathbf{d}^T =$ (d_{xx}, d_{xy}, d_{xz}, d_{yx}, d_{yy}, d_{yz}, d_{zx}, d_{zy}, d_{zz}). From Eq. 15, we know that given a diffusion-sensitive gradient $\mathbf{g}$, hence $\mathbf{b}$, the measured signal is the inner product of $\mathbf{b}$ and $\mathbf{d}$. Because the diffusion tensor is a symmetrical tensor, there are six independent diffusion coefficients. To solve for the six unknown variables, one needs to measure at least six attenuated signals by applying six different diffusion-sensitive gradients.

$$\ln[M_i(t)/M_0(t)] = -\mathbf{b}_i \cdot \mathbf{d}, \quad i = 1, 2, \ldots, N \text{ number of directions} \quad [16]$$

In the matrix form, we have

$$\begin{pmatrix} \ln\left(\frac{M_1}{M_0}\right) \\ \ln\left(\frac{M_2}{M_0}\right) \\ \vdots \\ \vdots \\ \ln\left(\frac{M_n}{M_0}\right) \end{pmatrix} = -\begin{pmatrix} b_1^T \\ b_2^T \\ \vdots \\ \vdots \\ b_n^T \end{pmatrix} \cdot \begin{pmatrix} d_{11} \\ d_{12} \\ d_{13} \\ d_{21} \\ d_{22} \\ d_{23} \\ d_{31} \\ d_{32} \\ d_{33} \end{pmatrix} \Rightarrow \mathbf{A} = -\mathbf{B} \cdot \mathbf{d} \quad [17]$$

Note that A, B, and d are, respectively, $n \times 1$, $n \times 9$, and 9×1 matrices. Diffusion coefficients can be uniquely determined by

$$\mathbf{d} = -\mathbf{B}^{-1} \cdot \mathbf{A} \quad [18]$$

where $\mathbf{B}^{-1}$ is the pseudo-inverse of the $\mathbf{B}$ matrix.

It is instructive to understand the relation between the diffusion gradient $\mathbf{g}$ and diffusion sensitivity $\mathbf{b}$ in waveforms. For the sake of convenience, let us consider it in one dimension. From Eq. 14, we know that b is just the time integral of q^2.

$$b(t) = 4\pi^2 \int_0^t q^2(\tau) d\tau = \int_0^t \gamma^2 g^2(\tau) \tau^2 d\tau \quad [19]$$

In the spin echo sequence with a continuous gradient, q changes sign at TE/2 because of the 180° RF pulse (Fig. 3). Therefore, the value of b is twice the integral of $\gamma^2 g^2(\tau)\tau^2$ from 0 to TE/2.

$$b = 2 \times \int_0^{TE/2} \gamma^2 g^2 \tau^2(\tau) d\tau = 2 \times \frac{1}{3}\gamma^2 g^2 \frac{1}{8} TE^3 = \frac{1}{12}\gamma^2 g^2 TE^3 \quad [20]$$

If we use a pair of pulsed gradients instead of a continuous gradient, the b value can be readily calculated by integrating the waveform of q^2 (Fig. 4).

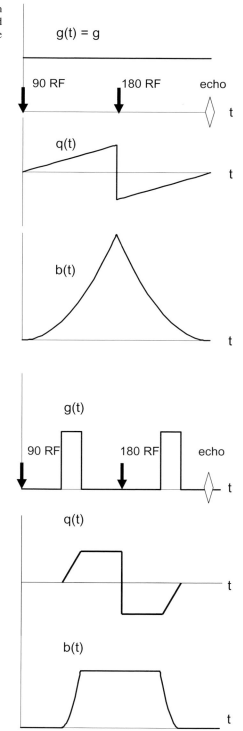

Fig. 3. Waveforms of the constant diffusion gradient g, the resulting $q(t) = \gamma \int g d\tau$, and $b(t) = \int_0^t \gamma^2 g^2 \tau^2 d\tau$ in the spin echo sequence

Fig. 4. Waveforms of the pulsed diffusion gradients $g(t)$, the resulting $q(t) = \gamma \int g(\tau) d\tau$, and $b(t) = \int_0^t \gamma^2 g^2(\tau) \tau^2 d\tau$ in the spin echo sequence

$$b = \frac{2}{3}\gamma^2 g^2 \delta^3 + \gamma^2 g^2 \delta^2 (\Delta - \delta) = \gamma^2 g^2 \left(\Delta - \frac{\delta}{3}\right) \quad [21]$$

This sequence, originally proposed by Stejskal and Tanner, is called a pulsed gradient spin echo (PGSE) sequence and is widely used in the current diffusion MRI experiment [9].

Diffusion Anisotropy and Diffusion Tensor Imaging

We have described the principles of diffusion measurement using MRI. In biological tissues, the diffusion is not isotropic; the diffusion coefficients in different directions may be different because the microsctructures restricting the diffusion may vary with the orientation. This orientation-dependent diffusion, called diffusion anisotropy, exists in most biological tissues; it was first demonstrated by Moseley et al. and Douek et al. [10, 11]. Diffusion tensor is a powerful metric to describe anisotropic diffusion in a biological tissue [12]. Given the diffusion tensor **D** of a tissue, the diffusion coefficient at any direction $\hat{u}$, D_{uu}, can be readily determined by

$$D_{uu} = \hat{u} \cdot \mathbf{D} \cdot \hat{u} \quad [22]$$

To understand Eq. 22 geometrically, we can picture the diffusion tensor as an ellipsoid with three axes of different lengths (Fig. 5). The three axes of the ellipsoid

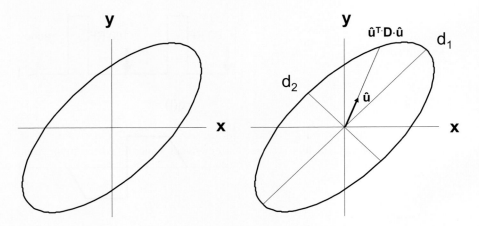

Fig. 5. Illustration of a diffusion tensor as an ellipse in the two-dimensional coordinates {x, y}. The diffusion coefficient in any direction $\hat{u}$ can be determined by the length from the origin to a point on the ellipse with the same angle as $\hat{u}$. The lengths in the major and minor axes are the first eigenvalue (d_1) and second (d_2) eigenvalue of the diffusion tensor, respectively

and their corresponding lengths can be determined by performing a mathematical procedure called "matrix diagonalization"; the resulting unit vectors of the axes and their lengths are, respectively, the eigenvectors and eigenvalues of the diffusion tensor. It is obvious that the diffusion coefficient in any direction û is determined by the length from the origin to a point on the ellipsoid surface with the same polar and azimuthal angles as û.

Having known eigenvalues of the diffusion tensor, we can calculate two important coordinate-invariant indices: the trace apparent diffusion coefficient (ADC) and fractional anisotropy [13]. Trace ADC, denoted $\langle d \rangle$, is quantified as the mean of the three eigenvalues of the diffusion tensor d_1, d_2, and d_3.

$$\langle d \rangle = (d_1 + d_2 + d_3)/3 \qquad [23]$$

Fractional anisotropy (FA) represents the degree of deviation of a diffusion ellipsoid from a sphere and is quantified as the standard deviation of the eigenvalues of the diffusion tensor normalized by the "magnitude" of the three eigenvalues of the diffusion tensor.

$$FA = \{3[(d_1 - \langle d \rangle)^2 + (d_2 - \langle d \rangle)^2 + (d_3 - \langle d \rangle)^2]/[2(d_1^2 + d_2^2 + d_3^2)]\}^{1/2} \qquad [24]$$

Many articles have reported the clinical values of $\langle d \rangle$ and FA; $\langle d \rangle$ was found to decrease within minutes following the onset of ischemic stroke, permitting early diagnosis of acute stroke. FA was considered to reflect the integrity of axonal myelination, and it was found to be decreased in the presence of demyelinating or dysmyelinating disease. Recently, Song et al. performed histological correlation and confirmed that demyelination leads primarily to increased diffusivity perpendicular to the axonal direction, whereas axonal injury results in decreased diffusivity along the axonal direction [14].

In addition to clinically relevant diffusion indices, researchers investigated the eigenvectors of the diffusion tensor and their correspondence with the underlying microstructure. In organized tissues such as cerebral white matter, skeletal muscle, and myocardium, the first eigenvector of the diffusion tensor was found to correspond to the local fiber orientation. Lin et al. investigated the first eigenvectors along the rat optic tracts by comparing them with the same optic tracts that were enhanced by manganese chloride [15]. They showed that the first eigenvectors aligned closely with the local orientations of the enhanced optic tracts. In the myocardium, it has been validated that the first eigenvector paralleled the myocardial fiber orientation, and that the third eigenvector corresponds to the normal myocardial sheet [16]. Because the first eigenvector of the diffusion tensor can be used to define local fiber orientation, it is sensible to streamline the eigenvectors, connect them, and display the connected paths as fiber tracts. This technique called diffusion tractography was first reported by Conturo et al. and Mori et al. in 1999 [17, 18]. It opens up the possibility of seeing white matter tracts of a human brain in vivo. To date, active research has been undertaken in this field exactly because diffusion MRI is the only in vivo imaging modality capable of showing white matter tracts.

Just as functional MRI allows us to investigate gray matter with regard to our cognition, diffusion MRI allows us to understand how different functional areas are connected via white matter tracts to form neural circuits.

Although diffusion tensor imaging (DTI) can provide information about diffusion properties and local fiber orientation, the information is inaccurate in regions having more than one fiber tract that cross each other. In the brain, there are many axonal fiber tracts running in three major directions: commissural fibers between the left and right hemispheres, association fibers in the anteroposterior direction, and projection fibers in the superoinferior direction. It is inevitable that these fiber tracts would constantly cross each other. Wiegell et al. showed that in regions in which fibers cross, such as the pons (where corticospinal tracts intersect with the middle cerebellar peduncle) and the centrum semiovale (where corticospinal tracts, corpus callosum, and superior longitudinal fasciculi cross), the first and second eigenvalues are almost equal and indistinguishable [19]. The shape of the diffusion tensor in these areas approximates a disk with its plane containing the first and second eigenvectors. Therefore, tractography based on DTI data is subject to erroneous tracing results.

NMR Signal and Probability Density Function of Water Molecules

To resolve the crossing fibers and to characterize more accurately the diffusion property and fiber orientation, let us review the basic theory of the MRI signal in the existence of diffusion. Consider a diffusion experiment using the pulsed gradient spin echo (PGSE) sequence with the gradient pulse duration δ and a time interval Δ between two gradient pulses. Also recall that the probability density function (PDF) $P_s(x_0|x, t)$ is the probability of finding the same molecule at position x at time t given the initial probability of unity at x_0. The pulsed gradients creates a phase shift to each spin due to translational displacement of diffusion, that is, $\Delta\phi = \gamma\delta \mathbf{g} \cdot (\mathbf{x}' - \mathbf{x})$. The echo signal is the summation of an ensemble of spin magnetizations.

$$S(\mathbf{g}) = \iint \rho(\mathbf{x}) P_s(\mathbf{x}|\mathbf{x}', \Delta) \text{Exp}(i\gamma\delta\mathbf{g} \cdot (\mathbf{x}' - \mathbf{x})) d\mathbf{x} d\mathbf{x}' \qquad [25]$$

$$S(\mathbf{q}) = \iint \rho(\mathbf{x}) P_s(\mathbf{x}|\mathbf{x}', \Delta) \text{Exp}(i\mathbf{q} \cdot (\mathbf{x}' - \mathbf{x})) d\mathbf{x} d\mathbf{x}' \qquad [26]$$

Replacing $\mathbf{x}$ with relative displacement $\mathbf{r} = \mathbf{x}' - \mathbf{x}$, we have

$$S(\mathbf{q}) = \iint \rho(\mathbf{x}) P_s(\mathbf{x}|\mathbf{x}+\mathbf{r}, \Delta) \text{Exp}(i\mathbf{q} \cdot \mathbf{r}) d\mathbf{x} d\mathbf{r} \qquad [27]$$

In a homogeneous microenvironment, relative displacement is independent of the initial position $\mathbf{x}$.

$$S(\mathbf{q}) = \left(\int \rho(\mathbf{x})d\mathbf{x}\right)\int P_s(\mathbf{r}, \Delta)\mathrm{Exp}(i\mathbf{q}\cdot\mathbf{r})d\mathbf{r}$$
$$= \text{constant} \times \int P_s(\mathbf{r}, \Delta)\mathrm{Exp}(i\mathbf{q}\cdot\mathbf{r})d\mathbf{r} \qquad [28]$$

Therefore, the MRI signal in the **q** space is the Fourier transform of PDF in the image space. In a heterogeneous microenvironment, the translational displacement **r** may vary with different positions **x**. We have

$$S(\mathbf{q}) = \int P_{av}(\mathbf{r}, \Delta)\mathrm{Exp}(i\mathbf{q}\cdot\mathbf{r})d\mathbf{r} \qquad [29]$$

where $P_{av}(\mathbf{r}, \Delta)$ is the weighting average of the PDF, called the average propagator.

$$P_{av}(\mathbf{r}, \Delta) = \int \rho(\mathbf{x})P_s(\mathbf{x}|\mathbf{x}+\mathbf{r}, \Delta)d\mathbf{x} \qquad [30]$$

Because most of the biological tissues are inhomogeneous and restrictive for molecular diffusion, Eq. 29 is generally applicable to brain imaging. In such conditions, we have a Fourier relation between the MRI signal $S(\mathbf{q})$ and the average propagator $P_{av}(\mathbf{r}, \Delta)$.

If we approximate $P_{av}(\mathbf{r}, \Delta)$ as a three-dimensional Gaussian, it can be formulated as a diffusion ellipsoid with the lengths of the major axes being d_1, d_2, and d_3.

$$P_{av}(\mathbf{r}, \Delta) = (4\pi\Delta)^{-3/2}(d_1 d_2 d_3)^{-1/2}\mathrm{Exp}[-\mathbf{r}\cdot\mathbf{D}^{-1}\cdot\mathbf{r}/(4\Delta)] \qquad [31]$$

According to Eq. 29, the MRI signal is the Fourier transform of Eq. 31

$$S(\mathbf{q}) = \mathrm{Exp}\left[-4\pi^2\int_0^t \mathbf{q}^T(\tau)\cdot\mathbf{D}\cdot\mathbf{q}(\tau)d\tau\right] \qquad [32]$$

Here we have an equation exactly the same as Eq. 12. Therefore, we know that DTI is based on the assumption that the average propagator can be approximated as a three-dimensional Gaussian. This assumption is valid in simple conditions with single-fiber orientations but is not valid in regions with complex fiber conditions such as fiber crossing, kissing, or splaying. To map complex fiber conditions, it is necessary to sample the MRI signal in the q-space and invoke Eq. 30 to obtain the average propagator without any approximation. This principle was first enunciated by Callaghan but was considered time-consuming because data in both the k-space and the q-space have to be sampled [20]. Wedeen et al. first implemented this principle in a human brain and successfully resolved crossing fibers in the pons and centrum semiovale [21]. As is discussed later, this technique requires hundreds of spectral data in the q-space for reconstructing the average propagator; hence, it is called diffusion spectrum imaging (DSI) [22].

To capture the idea of how DSI works and how it is different from DTI, it is instructive to illustrate its principle in two-dimensional schematics. As shown in

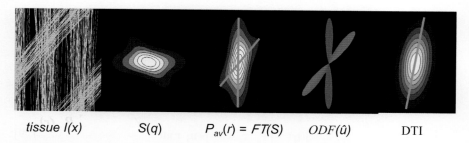

tissue I(x)   S(q)   P_{av}(r) = FT(S)   ODF(û)   DTI

Fig. 6. Illustration of crossing fibers I(x) within a pixel, acquired diffusion MRI signal in the q-space S(q), average propagator $P_{av}(\mathbf{r})$ after Fourier transform of S(q), and orientation distribution function $\mathrm{ODF}(\hat{\mathbf{u}}) = \int P_{av}(r\hat{\mathbf{u}})r^2 dr$. In contrast, the diffusion tensor is the first-order approximation of orientation distribution function (*ODF*), the first eigenvector of which cannot define the orientations of crossing fibers

Fig. 6, we have a pixel containing two fiber bundles running across each other I(**x**); the signal S(**q**) in this pixel will present different attenuation profiles along different directions in the q-space. Fourier transform of S(**q**) yields the average propagator $P_{av}(\mathbf{r})$. To express this in spherical coordinates without r dependence, $P_{av}(\mathbf{r})$ can be transformed to the orientation distribution function (ODF), $\mathrm{ODF}(\hat{\mathbf{u}}) = \int P_{av}(r\hat{\mathbf{u}})r^2 dr$. We can see that different orientations in crossing fibers are readily depicted by average propagator and ODF, whereas DTI can only define an eigenvector that is inaccurate for either one of the orientations. It should be noted that the average propagator or ODF only resolves different fiber orientations within a pixel; it cannot determine the location of the fibers in that pixel, nor can it differentiate fiber crossing from fiber kissing or splaying.

Diffusion Spectrum Imaging

Figure 7 illustrates the workflow of the DSI experiment. Hundreds of diffusion-weighted images are acquired. Each image is attenuated by a pair of diffusion gradients with a specific value of **q** corresponding to a grid point in the q-space. Having acquired the diffusion-weighted images, we have hundreds of sampled data in the q-space for each pixel. We perform a three-dimensional Fourier transform of the q-space data to obtain the average propagator in each pixel. To extract the fiber orientations, the ODF is calculated by taking the second moment of the average propagator along different radial directions. By defining the local maxima of the ODF, the fiber orientations are determined, color-coded, and mapped to each pixel.

Implementation of DSI requires estimation of optimum sampling parameters with the consideration of hardware performance. The ultimate goal is to obtain an

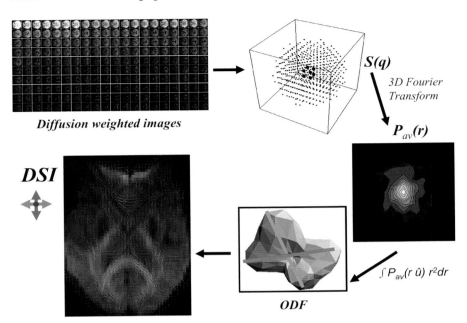

Fig. 7. Flow chart of diffusion spectrum imaging of human brain in a clinical scanner. *3D*, three-dimensional

average propagator that has orientation contrast. Knowing that the diffusion coefficient in the brain ranges from 0.5 to 2.5 $\mu m^2\,ms^{-1}$, the maximal b value of approximately 6000 s mm^{-2} would attenuate the MRI signal down to less than 5%. According to Eq. 21, it requires the maximal q value, q_{max}, of approximately 0.05 μm^{-1} provided the gradient duration is 30 ms, and the diffusion time is 70 ms. Furthermore, knowing that the diameter of an axon ranges from 2 to 20 μm, the field of view (FOV) of the average propagator should be at least twice the range, say ±50 μm. The sampling interval Δq is therefore 0.01 μm^{-1}. The maximal q and Δq estimated above require approximately $11^3 = 1331$ sample points in the q-space. By sampling the three-dimensional grid points inside the sphere with the radius of q_{max} and filling the unsampled points with zeros, the sampling number would reduce to 515. Compared with the standard DTI, with which only six samples are taken, DSI requires a substantially larger number of samples in the q-space (Fig. 8). Using echo planar imaging (EPI) with TR 6.5 s, the total scan time of DSI over the whole brain would take approximately 1 h.

The above estimation provides a rationale for the optimum Δq, q_{max}, and sampling numbers in the q-space. However, to implement DSI in the clinical setting, the scan time should be less than 15 min, roughly one-fourth of the estimated sampling number. There are several approaches to reducing the sampling number while keeping the orientation contrast unaffected. The first approach is to reduce the

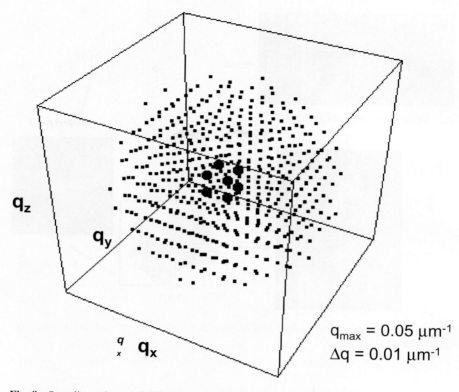

Fig. 8. Sampling schema of diffusion spectrum imaging (DSI). The sampled points are located at the three-dimensional Cartesian grid points in the q-space with the field of view $q_{max} = 0.05\,\mu m^{-1}$ and resolution $\Delta q = 0.01\,\mu m^{-1}$

sampling points by relaxing Δq or q_{max}. Either one has the potential risk of inducing errors. If Δq is broadened while keeping q_{max} unchanged, the FOV of the average propagator is decreased, which may cause aliasing. If q_{max} is reduced while keeping Δq unchanged, the MRI signal may not be attenuated sufficiently, which may cause a rippling artifact at the edge of the average propagator. To determine the optimum q_{max} and Δq for 515 and 203 sample points, we acquired highly sampled diffusion data with the b_{max} 9000 s/mm^2 and 925 diffusion gradients; we then subsampled at different values of b_{max} from 1000 to 7000 s/mm^2. Our results showed that the optimum b_{max} for 515 sampling points was close to 6000 s/mm^2, whereas it was approximately 3000 s/mm^2 for 203 sampling points [23].

The second approach is to down-sample the data in an interleaved sampling fashion and estimate the unsampled data by interpolating the adjacent sampled data or to employ the half-Fourier method by sampling data over a hemisphere and extrapolate the other hemisphere based on the conjugate symmetry of the q-space

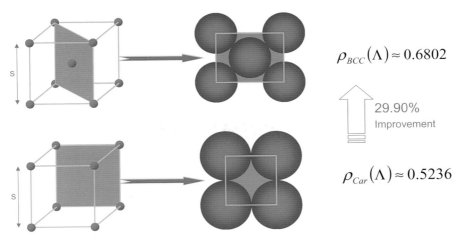

Fig. 9. Comparison of sampling efficiency between Cartesian (bottom) and body-center-cubic (BCC, top) sampling lattice. The packing density of BCC (ρ_{BCC}) is approximately 30% higher than that of the Cartesian packing density (ρ_{Car})

data [24, 25]. The third approach is to use a sampling scheme more efficient than the standard sampling at the Cartesian grid points. In crystallography, it is known that lattice points in the body–center–cubic (BCC) arrangement have higher packing density than the Cartesian grid points [24]. Under equal sampling density, the number of points of BCC sampling is estimated to be approximately 70% that of Cartesian sampling (Fig. 9). Furthermore, aliasing in the BCC sampling scheme appears at eight corners of a voxel cube. Compared with Cartesian sampling, it is less likely to overlap with the average propagator.

Another original approach, called q-ball imaging (QBI), is to acquire data on the surface of a sphere in the q-space, rather than sampling at the Cartesian grid points [26]. Fourier transform of data on a spherical surface, called the Funk-Radon transform, directly yields ODF. Because QBI requires fewer sampling points (typically <300 points) and lower gradient strengths (<3000 s/mm^2), it is easier than DSI to implement QBI in clinical scanners under current hardware performance.

Diffusion Tractography and Its Application

Diffusion tractography is to visualize white matter tracts by connecting coherent principal vectors obtained from diffusion MRI to form distinct fiber bundles. Based on this basic notion, software for tractography reconstruction has been developed and is widely available in the public domain. Although the algorithms may differ in details, they invariably include the key steps as follows: (1) select the seeds; (2) determine the tracking direction from the seeds; (3) set stopping criteria for tracking; and (4) use specific regions to filter the tracts of interest. Owing to the capability of resolving crossing fibers, the white matter tracts reconstructed from in vivo DSI data provide reliable morphological results even at the crossing regions [27]. For the main fiber tracts, such as the corpus callosum, pyramidal tracts, and superior longitudinal fasciculus, diffusion spectrum imaging (DSI) tractography shows satisfactory agreement with known anatomy (Fig. 10) [28].

Being able to see the white matter tracts, it is of great interest to see how they are related to neurological function. Consistent with the well-known language lateralization in right-handed men, Hagmann et al. found significant asymmetry of the arcuate fasciculus in this group of subjects [29]. Patients with brain tumors often present with variable neurological symptoms, from which we can correlate neurological deficits with tumor involvement of the functionally relevant tracts [30]. In a patient with recent onset of dyslexia, DSI tractography showed a hematoma in the left temporal lobe compressing the left arcuate fasciculus (Fig. 11). Another patient with gradual-onset left hemianopsia was found to have a tumor in the right temporooccipital lobe compressing the right optic tract (Fig. 12). In 15 patients with brain tumors and variable degrees of hemiparesis, we found that the severity of muscle power weakness had a moderate association with the displacement of corticospinal tracts (Fig. 13). Similar results were reported in patients with subcortical stroke in whom the distribution of extremity weakness was associated with the location and extent of the lesions [31, 32]. Owing to its potential clinical value, diffusion MRI has been actively used to discover subtle abnormalities in white matter tracts in various neuropsychiatric disorders, such as Alzheimer disease, schizophrenia, autism, obsessive-compulsive disorder, drug addition, and amyotrophic lateral sclerosis.

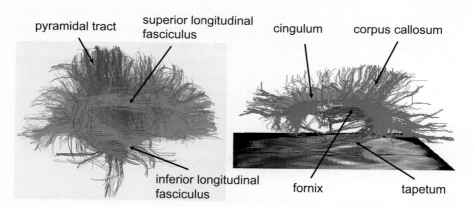

Fig. 10. Diffusion spectrum imaging tractography of a whole brain acquired with a 3-T MRI scanner. The morphology of individual tracts, color-coded according to their main orientations, is in agreement with known anatomy

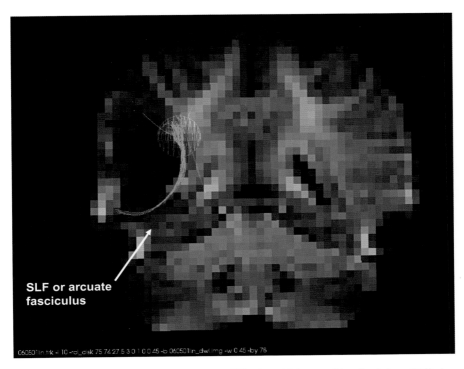

Fig. 11. A 49-year-old woman suffered from difficulty with her speaking for 6 days. Diffusion spectrum imaging tractography showed a hematoma in the left temporal lobe compressing the left arcuate fasciculus

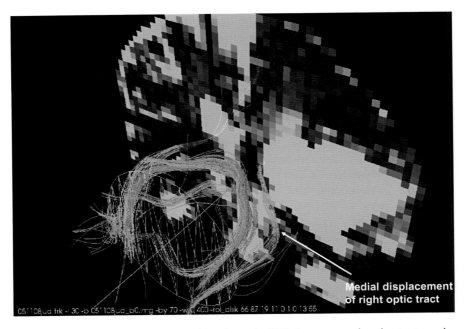

Fig. 12. A patient with gradual-onset left hemianopsia. Diffusion spectrum imaging tractography showed a right temporooccipital tumor compressing the right optic tract

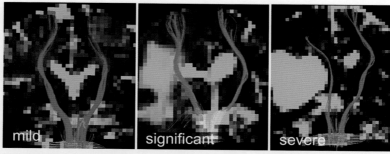

Muscle power Displacement	Normal (5)	Mild (4~5)	Severe (<4)
Mild	3	2	0
Significant	0	4	2
Severe	1	1	2

Fig. 13. Muscle power weakness and displacement of the corticospinal tract. The displacement was graded as mild, significant, or severe by the consensus of two experienced radiologists. Fisher's exact test based on two-point scoring showed moderate association between tract displacement and muscle power ($P = 0.08$)

Although diffusion tractography is a promising tool for studying the structural integrity of functional connectivity, further improvement is needed to bring this tool beyond visualization to become a quantitative analysis tool. With this tool, tractography can serve as a road map to sample the diffusion indices, such as trace ADC or fractional anisotropy, along the tracts of interest and obtain plots of diffusion indices along the tract coordinates. Normative data of such tract-specific diffusion indices must be established to characterize the difference in age, sex, and handedness so subtle abnormalities in neuropsychiatric disorders and their response to therapy can be assessed objectively.

In addition to establishing clinical values, the future directions for diffusion MRI research that are most important are to, first, develop novel techniques to "see" the tracts in action; second, combine other functional imaging modalities to define the neural circuits; and third, analyze the diffusion signature of the cortical cytoarchitecture.

Conclusion

Diffusion MRI allows us to see the morphology of white matter tracts and their structural integrity. This technique probes structural information at the microscopic level and maps it onto the whole brain. This unique capability opens a door to early diagnosis and better health care of patients with neuropsychiatric disorders and to the study of functional connectivity that underlies the coordination of brain function.

References

1. Turner R, Jones T (2003) Techniques for imaging neuroscience. Br Med Bull 65:3–20.
2. Harel N, Ugurbil K, Uludag K, et al (2006) Frontiers of brain mapping using MRI. J Magn Reson Imaging 23(6):945–957.
3. Mori S, Zhang J (2006) Principles of diffusion tensor imaging and its applications to basic neuroscience research. Neuron 51:527–539.
4. Paulson OB, Hertz MM, Bolwig TG, et al (1997) Filtration and diffusion of water across the blood-brain-barrier in man. Microvasc Res 13:113–124.
5. Nicholson C, Phillips JM (1981) Ion diffusion modified by tortuosity and volume fraction in the extracellular microenvironment of the rat cerebellum. J Physiol 321:211–223.
6. LeBihan D (1995) Molecular diffusion, tissue microdynamics and microstructure. NMR Biomed 8(7):375–386.
7. Hahn EL (1950) Spin echoes. Phys Rev 80:580–594.
8. Torrey HC (1956) Bloch equations with diffusion terms. J Chem Phys 104:563–565.
9. Stejskal EO, Tanner JE (1965) Spin diffusion measurement: spin echoes in the presence of a time-dependent field gradient. J Chem Phys 42(1):288–292.
10. Moseley ME, Cohen Y, Kucharczyk J, et al (1990) Diffusion-weighted MR imaging of anisotropic water diffusion in cat central nervous system. Radiology 176(2):439–445.
11. Douek P, Turner R, Pekar J, et al (1991) MR color mapping of myelin fiber orientation. J Comput Assist Tomogr 15(6):923–929.
12. Basser PJ, Mattiello J, LeBihan D (1994) Estimation of the effective self-diffusion tensor from the NMR spin echo. J Magn Reson B 103(3):247–254.
13. Basser PJ (1997) New histological and physiological stains derived from diffusion-tensor MR images. Ann N Y Acad Sci 820:123–138.
14. Song SK, Sun SW, Ramsbottom MJ, et al (2002) Dysmyelination revealed through MRI as increased radial (but unchanged axial) diffusion of water. Neuroimage 17(3):1429–1436.
15. Lin CP, Tseng WY, Cheng HC, et al (2001) Validation of diffusion tensor magnetic resonance axonal fiber imaging with registered manganese-enhanced optic tracts. Neuroimage 14(5):1035–1047.
16. Tseng WY, Wedeen VJ, Reese TG, et al (2003) Diffusion tensor MRI of myocardial fibers and sheets: correspondence with visible cut-face texture. J Magn Reson Imaging 17(1):31–42.
17. Conturo TE, Lori NF, Cull TS, et al (1999) Tracking neuronal fiber pathways in the living human brain. Proc Natl Acad Sci USA 96(18):10422–10427.
18. Mori S, Crain BJ, Chacko VP, et al (1999) Three-dimensional tracking of axonal projections in the brain by magnetic resonance imaging. Ann Neurol 45:265–269.
19. Wiegell MR, Larsson HB, Wedeen VJ (2000) Fiber crossing in human brain depicted with diffusion tensor MR imaging. Radiology 217(3):897–903.
20. Callaghan PT (1991) Principle of nuclear magnetic resonance microscopy. Clarendon Press, Oxford, pp 328–370.
21. Wedeen VJ, Reese TG, Tuch DS, et al (2000) Mapping fiber orientation spectra in cerebral white matter with Fourier-transform diffusion MRI. Proc Int Soc Magn Reson Med 8:82.
22. Wedeen VJ, Hagmann P, Tseng WY, et al (2005) Mapping complex tissue architecture with diffusion spectrum magnetic resonance imaging. Magn Reson Med 54(6):1377–1386.
23. Kuo LW, Wedeen VJ, Mishra C, et al (2007, in press) Determining optimum b maximum values for diffusion spectrum imaging and q-ball imaging in clinical MRI system. Proc Int Soc Magn Reson Med:312.

24. Chiang WY, Wedeen VJ, Kuo LW, et al (2006) Diffusion spectrum imaging using body-center-cubic sampling scheme. Proc Int Soc Magn Reson Med:1041.
25. Lin CP, Tseng WY, Weng JC, et al (2003) Reduced encoding of diffusion spectrum imaging with cross-term correction. In: 1st International IEEE EMBS Conference on Neural Engineering, Capri, Italy, pp 561–563.
26. Tuch DS (2004) Q-ball imaging. Magn Reson Med 52(6):1358–1372.
27. Wedeen VJ, Wang R, Benner T, et al (2005) DSI tractography of CNS fiber architecture and cortical architectonics. Proc Int Soc Magn Reson Med:584.
28. Kuo LW, Wedeen VJ, Weng JC, et al (2005) Reconstruction and visualization of white matter tracts based on clinical diffusion spectrum imaging. Proc Int Soc Magn Reson Med:1313.
29. Hagmann P, Cammoun L, Martuzzi R, et al (2006) Hand preference and sex shape the architecture of language networks. Hum Brain Mapp 27(10):828–835.
30. Brian JJ, Aaron SF, Joshua M, et al (2004) Diffusion tensor imaging of cerebral white matter: a pictorial review of physics, fiber tract anatomy, and tumor imaging patterns. AJNR Am J Neuroradiol 25(3):356–369.
31. Lie C, Hirsch JG, Rossmanith C, et al (2004) Clinicotopographical correlation of corticospinal tract stroke: a color-coded diffusion tensor imaging study. Stroke 35(1):86–92.
32. Lee JS, Han MK, Kim SH, et al (2005) Fiber tracking by diffusion tensor imaging in corticospinal tract stroke: topographical correlation with clinical symptoms. Neuroimage 26(3):771–776.

Overview of MR Diffusion Tensor Imaging and Spatially Normalized FDG-PET for Diffuse Axonal Injury Patients with Cognitive Impairments

Ayumi Okumura, Jun Shinoda, and Jituhiro Yamada

Summary

We detected and compared brain dysfunction areas using both magnetic resonance (MR) diffusion tensor imaging (DTI) and spatially normalized ^{18}F-fluorodeoxyglucose positron emission tomography (^{18}F-FDG-PET; statistical FDG-PET) for the patients with memory and cognitive impairments due to traumatic brain injury. A total of 70 diffuse axonal injury patients with memory and cognitive impairments (DAI patients) and 69 age-matched normal subjects were studied with DTI and statistical FDG-PET. All subjects underwent examinations with a 1.5-T Signa MRI system. We used a single-shot spin-echo echoplanar sequence for diffusion-tensor analysis. Tractographic results were analyzed with diffusion tensor visualization (dTV) software. The group analysis for fractional anisotropy (FA) was performed with the SPM99 software package. FDG-PET was performed on all patients. After PET imaging, statistical analysis using the easy Z-score imaging system (eZIS) was undertaken with processing steps that included smoothing, normalization, and Z transformation with respect to a normal database. The Z-score map was superimposed on a three-dimensional MRI brain scan. Group analysis was performed using statistical parametric mapping analysis. A decline of FA was observed around the corpus callosum in the DAI patients compared with that of normal subjects, and reduced glucose metabolism was seen in the cingulate association. These results suggest that the reduced metabolism in the cingulate cortex indicated deprived neuronal activation caused by the impaired neuronal connectivity that was revealed with DTI. Furthermore, the metabolic abnormalities in the cingulate cortex may be responsible for memory and cognitive impairments. Age-related cerebral metabolic and blood flow decline was observed in the anterior cingulate association. The clinical combination of DTI and statistical PET has a role in

---

Department of Neurosurgery, Chubu Medical Center for Prolonged Traumatic Brain Dysfunction, Kizawa Memorial Hospital, 630 Shimokobi, Kobi-cho, Minokamo, Gifu 505-0034, Japan

neuroimage interpretation for patients with memory and cognition impairments because its three-dimensional visualization allows more objective and systematic investigation.

Key words Diffuse axonal injury, Diffusion tensor imaging, eZIS, Tractography, FDG-PET SPM

Introduction

Cognitive and vocational sequelae are frequent complications after traumatic brain injury without obvious neuroradiological lesions. They may present as memory disturbance, impaired multitask execution, and loss of self-awareness. These symptoms have been attributed to diffuse brain injury and the diffuse loss of white matter or neural networks in the brain. There is currently no accurate method to diagnose and assess the distribution and severity of diffuse axonal injury. As computed tomography (CT) and magnetic resonance imaging (MRI) findings underestimate the extent of diffuse axonal injury and correlate poorly with the final neuropsychological outcome, this dysfunction tends to be clinically underdiagnosed or overlooked. Indirect evidence for loss of functional connectivity after nonmissile traumatic brain injury (nmTBI) has been provided by functional neuroimaging, such as positron emission tomography (PET). Most functional neuroimaging studies conducted after nmTBI have demonstrated that cognitive and behavioral disorders are correlated with some degree of secondary hypometabolism or hypoperfusion in regions of the cortex. To date, however, there has been no direct, in vivo demonstration of structural disconnections in patients with nmTBI who do not have macroscopically detectable lesions.

Diffusion tensor MRI (DTI), which measures diffusion anisotropy in vivo, is a promising method for noninvasively identifying the degree of fiber damage in various disease processes affecting the white matter. In biological systems, the diffusional motion of water is impeded by tissue structures such as cell membranes, myelin sheaths, intracellular microtubules, and associated proteins; motion parallel to axons or myelin sheaths is inhibited to a lesser degree than perpendicular motion, a phenomenon known as diffusion anisotropy. To reduce interindividual variability and to evaluate the whole brain objectively, a statistical normalizing method, MRI voxel-based analysis, has been developed.

We investigated which regions in the whole brain are commonly injured in nmTBI patients with cognitive impairments but no macroscopic lesions using voxel-based analysis of fractional anisotropy (FA), referred to as diffusion anisotropy. The advent of DTI has allowed interregional fiber tracking, called MR tractography, which reconstructs the three-dimensional trajectories of white matter tracts. The second purpose of our study was to investigate the clinical applicability of DTI in combination with spatially normalized ^{18}F-fluorodeoxyglucose positron emission tomography (FDG-PET).

Materials and Methods

Patient Population

We studied 70 patients in the chronic stage after they had suffered severe nmTBI and had recovered from their coma. All had sustained a high-velocity, high-impact injury in a motor vehicle accident. Their clinical features and the results of neuropsychological examinations are presented in Table 1. The study population was selected from among 157 consecutive nmTBI patients who entered the rehabilitation program of Chubu Medical Center. Patients with severe language or attention deficits that prevented neuropsychological testing were excluded from the study. Also excluded were patients with physiological deficits or neuroradiologically detectable lesions >1.6 cm^3, such as contusions, hematomas, or infarcts. Neuropsychological testing was performed within 2 weeks of obtaining MRI scans.

For the overall estimation of their global intellectual, mnemonic performance and attention, we administered the Wechsler Adult Intelligence Scale–Revised (WAIS-R) test, the Mini-Mental State Examination (MMSE), the Wechsler Memory Scale–Revised (WSM-R) test, and the Paced Auditory Serial Addition test (PASAT). For each of these tests, an adjusted performance score below the lower limit of the 95% tolerance interval of the normal population was considered indicative of pathology in this study. The controls were 69 sex- and age-matched normal subjects. Neither the patients nor the control subjects had a history of neurological or psychiatric disorders. All participants gave prior written informed consent. The protocol was approved by the Research Committee of the Kizawa Memorial Hospital Foundation.

MRI Scanning Protocol

All subjects underwent examinations with a 1.5-T Signa MRI system (GE Medical Systems, Milwaukee, WI, USA). We used a single-shot spin-echo planar sequence (TR/TE 10000/79 ms, slice thickness 3 mm, field of view 25 cm^2, number of experiments 4, pixel matrix 128 × 128 pixels) for diffusion tensor analysis. Diffusion gradients (b = 1000 s/mm^2) were always applied on two axes simultaneously around the 180 pulses. Diffusion properties were measured along six noncollinear directions. Diffusion-weighted MR images were transferred to a workstation supplied by the manufacturer (Advantage Workstation, GE Medical Systems); structural distortion induced by large diffusion gradients was corrected based on T2-weighed echo-planar images (b = 0 s/mm^2). The six elements of the diffusion tensor were estimated in each voxel assuming a monoexponential relation between signal intensity and the b-matrix. Using multivariate regression analysis, the eigenvectors and eigenvalues ($\lambda_1 > \lambda_2 > \lambda_3$) of the diffusion tensor were determined. The FA maps were generated on a voxel-by-voxel basis as follows.

$$FA = \sqrt{3/2} \times \sqrt{[(\lambda_1 - MD)^2 + (\lambda_2 - MD)^2 + (\lambda_3 - MD)^2]} / \sqrt{(\lambda_1^2 + \lambda_2^2 + \lambda_3^2)}$$

Spatial Normalization and Voxel-Based Analysis Using the FA Map

Spatial normalization is an essential preprocessing step in voxel-based analysis. As the contrast of the FA map is different from that of T1- or T2- weighted template images provided with the statistical parametric mapping (SPM99) software package (Wellcome, Department of Cognitive Neurology, London, UK), we created the FA template from a control group of 10 normal subjects. T2-weighed echo-planar images of the normal controls were segmented into gray and white matter and cerebrospinal fluid using SPM99 software running in MATLAB 5.3 (Mathworks, Natick, MA, USA). The segmented white matter images in native space were transferred to the white matter templates using the residual sum of squared differences as the matching criterion. The parameter of the transformation was also applied to the corresponding FA map. The normalized FA maps were smoothed with an 8-mm full-width at half-maximum (FWHM) isotropic Gaussian kernel, and a mean image (FA template) was created. All FA maps in native space were then transferred onto the stereotactic space by registering each of the images to the FA template image. Normalized data were smoothed with an FWHM isotropic Gaussian kernel. Statistically significant differences between the patients and controls were tested with the SPM99 software package without global normalization. To test our hypotheses with respect to regionally specific group effects, we compared the estimates using two linear contrasts (increased or decreased FA in patients versus controls). The threshold was set to uncorrected $p < 0.001$; significance levels were set at corrected $p < 0.05$.

MR Tractography

Magnetic resonance (MR) tractography was made with diffusion tensor visualization software. For tractography of the corpus callosum, seed-volumes were located from the genu to the splenium through the body on reconstructed midsagittal images. For tractography of the fornix, seed-volumes were located in the column of the fornix on reconstructed coronal images. The FA value for stop criteria was 0.18. Tractographic results were overlaid on T2-weighted images.

PET Examination

The PET scanner used in this study was an ADVANCE NXi Imaging System (General Electric Yokokawa Medical Systems, Tokyo, Japan), which provides 35 transaxial images at 4.25-mm intervals. The in-plane spatial resolution (FWHM) was 4.8 mm. The scan mode was the standard two-dimensional mode. Subjects were placed in the PET scanner so the slices were parallel to the canthomeatal line. Immobility was checked by alignment of three laser beams, with lines drawn on the patient's face. Subjects fasted for at least 4 h prior to injection of FDG. An FDG dose of 0.12 mCi/kg was injected intravenously through the cubital vein over a period of 1 min. The subjects were comfortably seated with their eyes open and environmental noises kept to a minimum for approximately 40 min. After a 40-min rest, a ^{68}Ge/^{68}Ga rotating pin source was used to obtain 3-min transmission scans. Static PET scan was performed continuously for 7 min. A static scan reconstruction was performed with corrections for photon attenuation using data from the transmission scans, dead time, random, and scatter.

Images were processed and analyzed on a Microsoft workstation using MATLAB software (Mathworks, Sherborn, MA, USA), an easy Z-score imaging system (eZIS), and SPM99 software (courtesy of the Functional Imaging Laboratory, Wellcome Department of Cognitive Neurology, University College, London, UK).

Results

Table 1 shows the demographics of the traumatic brain injury patients and normal subjects. Patients in this study show almost 80% performance of memory and cognition compared with normal subjects.

Table 1. Demographic of traumatic brain injury patients and normal subjects

Factor	Patients	Normal
Mean age (years)	27.43 ± 12.09	28.23 ± 11.90
Sex		
Male	19	19
Female	4	4
Duration of impaired consciousness (days), mean ± SD	7.2 ± 8.6	—
Injury to MRI interval (months), mean ± SD	14.16 ± 16.2	—
WAIS-R		
Full scale IQ	80.4 ± 11.2	104.2 ± 14.1
Verbal IQ	84.6 ± 12.2	103.6 ± 12.9
Performance IQ	78.0 ± 13.5	105.2 ± 15.9
MMSE	26.3 ± 4.5	29.7 ± 0.8
WMS-R		
General memory	71.7 ± 14.6	106.3 ± 12.1
Delayed memory	66.5 ± 13.2	103.7 ± 9.8
Visual memory	71.0 ± 14.0	102.1 ± 7.5
Attention	85.3 ± 16.3	107.3 ± 7.2
PASAT	34.5 ± 10.6	46 ± 6.2

MRI, magnetic resonance imaging; IQ, intelligence quotient; WAIS–R, Wechsler Adult Intelligence Scale–Revised; MMSE, Mini-Mental State Examination; WMS–R, Wechsler Memory Scale–Revised; PASAT, Paced Auditory Serial Addition Test

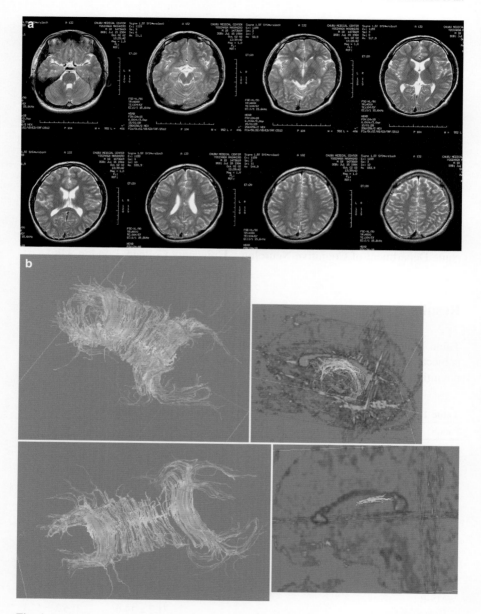

Fig. 1. a Conventional T2-weighted image (WI) of magnetic resonance imaging (MRI) of a traumatic brain injury patient (case 1) with cognitive impairment. There is no apparent abnormal lesion in this image. **b** Tractography of the corpus callosum and fornix of this patient compared to those of normal subjects

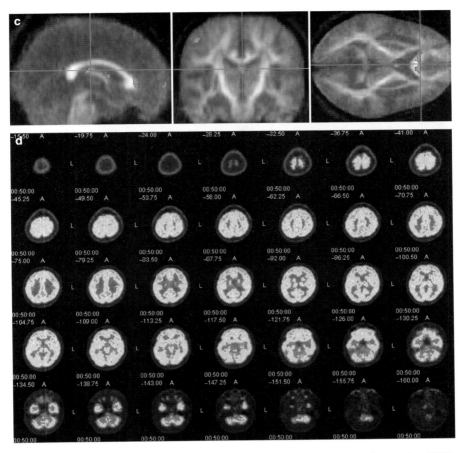

Fig. 1. c Fractional anisotropy (FA) reduction area on FA statistical parametric mapping (SPM) analysis. Red pixels show reduction of FA compared with normal. d Conventional fluorodeoxyglucose positron emission tomography (FDG-PET) of case 1. These images cannot display any metabolic abnormality

Figures 1 and 2 compare MR tractography of the corpus callosum and fornix of cases 1 and 2 (chosen from among the nmTBI cases) with that of normal subjects. In the nmTBI cases, the tracking lines through the genu and the body of the corpus callosum were thin and different from those of normal subjects, and the connecting fibers did not reach the frontal lobe. The volume and the connecting fibers from the splenium in the nmTBI patients were relatively retained. Compared to normal subjects, the nmTBI scans showed that the tracking lines through the column of the fornix were thin and did not pass along the fimbria of the hippocampus, although tractography around the mamillary body to the column of the fornix was relatively retained. An area of FA reduction was also seen around the corpus callosum and fornix (Fig. 1c).

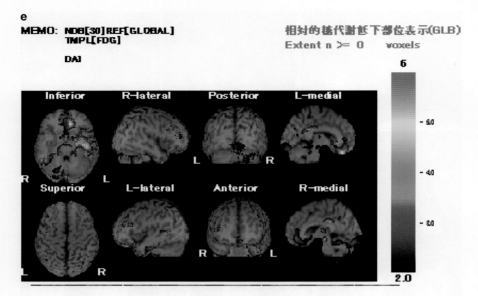

Fig. 1. e Easy Z-score imaging system (eZIS) reveals cerebral metabolic reduction in case 1. The areas of statistically significant reduction of metabolism are superimposed on brain surface images of each hemisphere. The color scale indicates the degree of significance (red > green > blue). This map shows abnormal metabolic voxels in posterior cingulate cortex

The eZIS views show cerebral metabolic reduction in these cases. Areas of statistically significantly reduced metabolism are superimposed on brain surface images of each hemisphere. The color scale indicates the degree of significance (red > green > blue). This map shows abnormal metabolic voxels in the posterior cingulate cortex.

In the group analysis, statistical parametric mapping (SPM) analysis indicated that the common area of FA reduction was the middle of the corpus callosum (Fig. 3). SPM analysis revealed the presence of significant hypometabolism in the medial prefrontal region, the medial frontobasal region, the anterior and posterior regions of the cingulate gyrus, and the thalamus bilaterally in all patients (Fig. 4) when compared with that of normal control subjects.

Discussion

Our results suggest that DTI was able to demonstrate objectively the abnormalities in nmTBI patients with cognitive impairments but no macroscopically detectable lesions. Voxel-based FA analysis objectively demonstrated the vulnerability of the corpus callosum and the fornix in nmTBI patients. The parasagittal subcortical white matter, internal capsules, cerebellar folia dorsal to the dentate nuclei, and the

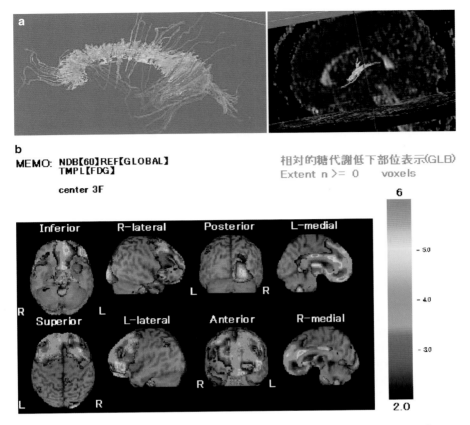

Fig. 2. a Tractography from the corpus callosum of case 2 patient, a 51-year-old man who was injured 8 months ago. Fibers from the corpus callosum of this case are coarse compared to those of a normal subject. **b** eZIS views of cerebral metabolic reduction in case 2. Areas of statistically significant reduced metabolism are superimposed on brain surface images of each hemisphere. The color scale indicates the degree of significance (red > green > blue). This map shows abnormal metabolic voxels in the anterior cingulate cortex

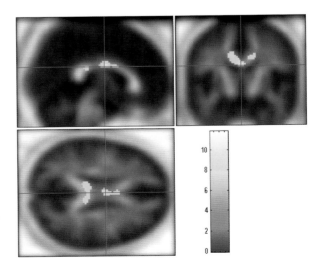

Fig. 3. SPM-rendered brain images of FA in normal subjects versus those of the diffuse axonal injury group

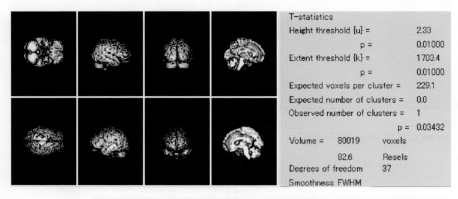

Fig. 4. SPM-rendered brain images of FDG-PET in normal subjects versus the diffuse axonal injury group. The orange areas represent statistical significance for a reduction in FDG in diffuse axonal injury group compared with controls. The spatial extent of orange cluster in voxels is located around the mesiofrontal cortex including the anterior cingulate association

brain stem, but not the corpus callosum or the fornix, are known to be susceptible to diffuse axonal injury. Our study was able to disclose the specific vulnerability of the corpus callosum and the fornix because our nmTBI patients had only cognitive impairments (no physiological problems).

The corpus callosum and the fornix are thought to be the structures at the core of the neural networks in cognition and memory. Changes in the anterior white matter, including the genu of the corpus callosum, are strongly related to age-related cognitive decline. The fornix is the major limbic white matter pathway interconnecting the hippocampus and the mamillary bodies. The limbic circuitry of the hippocampus–fornix–mamillary body interaction has been the focus of extensive research in memory function. Damage to any of these limbic structures results in various memory disorders.

The cingulate gyrus is a principal component of the limbic system, and its anterior and posterior parts have different thalamic and cortical connections and different cytoarchitecture, and they serve distinctive functions. The anterior cingulate gyrus can be divided into discrete anatomical and behavioral subdivisions: the affective division and the cognitive division. The affective division includes the areas of Brodmann 25 and 33 and the rostal area of Brodmann 24, and it plays a role in emotion and motivation; the cognitive division includes caudal areas of Brodmann 24 and 32 and plays a role in complex cognitive/attentional processing. Previous research indicates that significant regions of age-related decline remain prominent in the prefrontal cortex as well as in the anterior cingulate gyrus.

The posterior cingulate gyrus plays a part in orientation within and interpretation of the environment, and it has connections and behavioral attributes distinct from those of the anterior cingulate gyrus. Therefore, it is likely that the functions of these divisions are coordinated. The posterior cingulate gyrus also has dense connections with the medial temporal memory system. These communications are likely to contribute to the role of the posterior cingulate gyrus in orientation. The posterior cingulate gyrus is a major locus that is functionally involved early in the course of Alzheimer's disease. In this study, glucose hypometabolism in DAI patients existed in the whole of the cingulate gyrus as if the phenomena of both aging and Alzheimer's disease occurred at the same time.

The thalamus is also a principal component of the limbic system and the ascending reticular activating system. Therefore, the thalamus has distinct connections with the cingulate gyrus, reticular formation, and cerebral cortex.

There are abundant anatomical connections between the medial parietal region (precuneus)/posterior cingulate and the medial prefrontal region (medial frontal region)/anterior cingulate. These regions are functionally integrated in reflective self-awareness and the resting conscious state.

Axonal damage in the white matter, corpus callosum, and region of the superior cerebellar peduncle is common in patients with DAI. Fibers of the corpus callosum send axon collaterals to the cingulate gyrus. In the present study, glucose hypometabolism was demonstrated bilaterally in the medial prefrontal regions, medial frontobasal regions, cingulate gyrus, and thalamus in every patient with DAI. These results may provide strong evidence that white matter tract disruption, which leads to cortical disconnection resulting in cognitive decline and consciousness disturbance, occurs in patients with DAI.

Acknowledgments The authors thank Seisuke Fukuyama, Yukinori Kasuya, and Ryuji Okumura (Kizawa Memorial Hospital) and Naoki Hirata (General Electric Yokogawa Medical Systems, Tokyo, Japan) for their technical assistance.

References

1. Nakayama N, Okumura A, Shinoda J, et al (2006) Evidence for white matter disruption in traumatic brain injury without macroscopic lesions. J Neurol Neurosurg Psychiatry 77: 850–855.
2. Okumura A, Yasokawa Y, Nakayama N, et al (2005) The clinical utility of MR diffusion tensor imaging and spatially normalized PET to evaluate traumatic brain injury patients with memory and cognitive impairments. No To Shinkei 57:115–122.
3. Miwa K, Shinoda J, Yano H, et al (2004) Discrepancy between lesion distributions on methionine PET and MR images in patients with glioblastoma multiforme: insight from a PET and MR fusion image study. J Neurol Neurosurg Psychiatry 75:1457–1462.
4. Soeda A, Nakashima T, Okumura A, et al (2005) Cognitive impairment after traumatic brain injury: a functional magnetic resonance imaging study using the Stroop task. Neuroradiology 47:501–506.
5. Nakayama N, Okumura A, Shinoda J, et al (2006) Relationship between regional cerebral metabolism and consciousness disturbance in traumatic diffuse brain injury without large focal lesions: an FDG-PET study with statistical parametric mapping analysis. J Neurol Neurosurg Psychiatry 77:856–862.
6. Yasokawa Y, Shinoda J, Okumura A, et al (2007) Correlation between diffusion-tensor magnetic resonance imaging and motor evoked potential in chronic severe diffuse axonal injury. J Neurotrauma 24:163–173.

Transcranial Magnetic Stimulation in Cognitive Brain Research

S-H. Annabel Chen

Summary

Transcranial magnetic stimulation (TMS) is a widely used technique for noninvasive study of basic neurophysiological processes and the relation between the brain and behavior. It has a unique contribution in that it helps determine the need for brain areas for a particular task without the problem of diaschisis and compensatory plasticity seen with traditional lesion studies and that correlation methods of functional neuroimaging alone lack. In cognitive brain research, TMS has been employed to study perception, attention, learning, plasticity, language, and awareness. Its utility has also been extended to therapeutic research on neuropsychiatric conditions including mood disorders, schizophrenia, obsessive-compulsive disorders, and movement disorders. TMS can be combined with other neuroimaging techniques, such as electroencephalography, positron emission tomography, and functional magnetic resonance imaging to evaluate systematically the functional contribution of specific brain regions to cognitive task performance. A brief overview of the current contributions of TMS in cognitive brain research is presented, and important issues to be resolved in the application of TMS to cognitive studies are highlighted. The persistence of the TMS effect and the safety of this method for human studies are also briefly discussed.

Key words Transcranial magnetic stimulation, Cognition, Functional neuroimaging, Neuromodulation, Neuroplasticity

Introduction

Transcranial magnetic stimulation (TMS) is a method used to activate the brain by modulating the voltage over the membranes of cortical neurons. It is done noninvasively and is, by and large, painless, allowing researchers to stimulate discrete

---

Department of Psychology, Neurobiology and Cognitive Science Center, National Taiwan University, 1, Section 4, Roosevelt Road, Taipei 106, Taiwan

brain areas. This technique of brain stimulation has been available since the mid-1980s [1]. TMS uses Faraday's principles of electromagnetic induction, where short current pulses are driven through a coil, generating a brief magnetic field. A secondary current is induced in any nearby conductor if the magnitude of this magnetic field changes over time, and the size of the current is determined by the rate of change of the field. In human studies using TMS, the brain becomes the nearby conductor. Thus, it has been suggested that the term TMS is a misnomer for TMIES (transcranial magnetically induced electrical stimulation) [2].

How, then, does TMS work to effect a change in our brains? When a TMS coil is placed against the person's scalp, the current passing through it induces an electrical field in the brain affecting the transmembrane potential, which could lead to local membrane depolarization and firing of the neuron. At the neurophysiological and behavioral level, this may translate into evoked neuronal activity [on electroencephalography (EEG)]; changes in blood flow and metabolism [seen by positron emission tomography (PET), functional magnetic resonance imaging (fMRI), near-infrared spectroscopy (NIRS), or single photon emission computed tomography (SPECT)]; muscle twitches [seen by electromyography (EMG)]; or changes in behavior (revealed by cognitive effects and phosphenes). Thus, it can be considered a method for neuromodulation or, simply put, brain manipulation.

TMS Components

The duration of TMS stimulation required depends on the type of TMS stimulator used. A commercially available TMS machine operates by charging one or more capacitors within a matter of seconds and discharging the stored energy at high voltage into the coil as a single pulse (SpTMS) or as several repeating pulses (rTMS). A single-pulse stimulator machine usually discharges a current through the coil to produce a monophasic magnetic field, peaking at about 100 μs and decaying within 1.0 ms, which then induces an almost monophasic electrical field that is over in 200 μs. Monophasic current flow is considered most accurate, with lower heat and noise; however, because of the time interval required to recharged it to produce a second pulse, it is not easy to obtain bilateral cortical responses. A more sophisticated rapid-pulse TMS machine is capable of presenting pulses up to a frequency of 30–100 Hz for durations of more than 1 s. Such a stimulator requires changes in the circuitry, wherein the resistor is omitted and the diode is positioned across the thyristor; and the magnetic field and induced electrical field are then biphasic [2]. This setup allows approximately half of the original stored energy in the capacitors to be restored, enabling them to be recharged swiftly. In other words, the biphasic waveform of the induced electrical field allows lower currents to be used, shortening the recharging time. In comparison with the single-pulse stimulator, the rapid-pulse stimulator is able to disrupt neuronal functioning for a longer period of time with a duration of repetitive TMS pulses, whereas the single-pulse stimulator has an effect for only tens of milliseconds, as several seconds are needed before the next pulse can be delivered. However, biphasic pulses tend to produce higher noise and heat levels and may be less accurate than monophasic pulses.

Apart from the type of TMS machine used, as mentioned above, the depth and coverage of stimulation generated by a TMS pulse is dependent on the type of coil used. With a simple round TMS coil (as shown in Fig. 1a) an electrical field is induced in the tissue underneath the coil by the magnetic field around the coil created

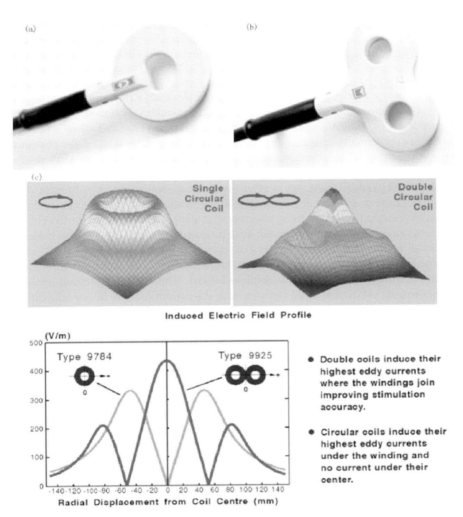

Fig. 1. Transcranial magnetic stimulation (TMS) coils and corresponding electric field distributions. **a** Simple circular coil: Magstim 90-mm coil. **b** Double coil: Magstim 70-mm coil. **c** Three-dimensional induced electric fields shown for circular 90-mm (*left*) and double 70-mm (*right*) coils. The circular coil has a larger area of effect, with the induced field being zero directly under its center. The double 70-mm coil has a more focal area of effect. (From Jalinous [3], with permission)

by the current flow. It is important to note that the induced current in the brain tissue is opposite to that in the coil, with a concentric electrical field being weakest in the center of the coil, and there is no single "hot spot" where the field is stronger than anywhere else (Fig. 1c) [3]. Based on this simple round TMS coil, it is therefore misleading to refer to TMS as "focal" as it cannot be brought to a focus. However, depending on the placement of the coil, there are selective local effects on the brain tissue underneath it. For example, a shift in position of as little as 1 cm over the visual cortex could lead to a perceived change in position, intensity, and quality of a visual phosphene [4]. Barker [5] showed that coil orientation significantly affects the strength of the stimulation. He found that even slight changes in coil position on the scalp can alter the locus of effective stimulation. This is because fibers can be depolarized with TMS only by a change in the electrical field across cell membranes. These considerations are probably trivial if we are stimulating cortex fibers tangential to the coil. However, there are some long, relatively unconvoluted white matter tracts underneath superficial cortex—such as the visual radiations from the dorsal lateral geniculate nucleus (dLGN) to the striate cortex (V1) and the compact projections from V1 to several surrounding extrastriate areas—that are sufficiently close to the cortical surface to be affected by the TMS. Here the coil orientation is an important consideration, yet is often overlooked. Sometimes it is thought that positioning the single coil at right angles to the skull, where only the rim rests on the scalp, would obtain the most effective stimulation. However, when the coil is so positioned, most of the magnetic field is outside the cortex, and the induced electrical field is minimal and ineffective, even at the point of contact [2].

The figure-of-eight coil, or butterfly or double coil (Fig. 1b), has been introduced to reduce several of the problems mentioned above. This coil configuration consists of two round coils mounted side by side with the current rotating in opposite directions in the two coils. The strongest electrical field is therefore induced in the tissue underneath the junction of these two coils, and the field is contiguous, being about twice as strong underneath the center than at any other part of the coils (Fig. 1c). However, to obtain this effect, the double coil needs to be placed roughly symmetrically with respect to the skull. If it is tilted such that one coil is closer to the skull than the other, the latter's magnetic field is too far from the brain to induce a sufficiently high electrical field in the tissue, and the strongest electrical field is below one of the coils, not at the juncture. One way to avoid this is to angle the two coils toward each other to approximate the contours of the skull. In addition, much smaller coils produce better localization but with weaker magnetic fields and may be inadequate for exciting a sufficiently large volume of cortex. An 8-cm double coil provides concentrated stimulation that is effective down to approximately 2 cm below its center, which is sufficient for most cognitive experiments [5].

Use of TMS in Cognitive Studies

There has been a rapid increase in investigations employing the TMS technique during the last decade [6]. In human studies, TMS has been applied to investigate sensation, perception, voluntary movements, awareness, language, memory, and problem solving. It has also been applied for diagnostic or therapeutic investigations in various clinical populations, such as neurological patients with neglect, stroke, Parkinson's disease, dementia, tinnitus, neuromuscular disease, epilepsy, brain tumors, and psychiatric patients with mood and psychotic disorders. Therefore, use of TMS in neuromodulation can be largely divided into three groups: (1) investigations of cognitive functions as a means of noninvasively affecting a targeted function or neurocircuitry; (2) diagnostic aid for presurgical planning and fiber track integrity; and (3) a therapeutic tool for managing illness. The current chapter focuses mainly on the first group of applications, in particular in the field of psychology.

In studies of cognitive function, Jahanshahi and Rothwell [7] summarized some important questions that can be addressed by TMS. As TMS produces a temporary lesion effect, it can determine whether the contribution of a target area is essential for performance. It can also investigate the timing involvement of a target area, that is, the time window during which the contribution of an area is essential for performance. In terms of temporal resolution of cognitive processing, it can determine the relative timing of the contribution of two or more areas to task performance. TMS has been employed to study neuroplasticity in terms of the changes in excitability in target areas with learning. Intracortical and transcallosal connectivity can be studied with TMS by determining what the subthreshold stimulation effect of one area has on subsequent suprathreshold stimulation of another target site. Medication effects on changes in cortical excitability in target areas and its time course can also be studied using TMS. Using TMS to measure cortical excitability can help understand the neuroplasticity of target areas following interventions such as neurosurgery or cognitive rehabilitation programs as well as after brain damage and neurological and psychiatric illness.

Common TMS Paradigms Used in Cognitive Studies

The noninvasive characteristics of TMS potentially allow it to be employed to investigate a broad variety of issues in cognitive neuroscience. We can design experimental protocols to address questions concerning location, timing, lateralization, functional relevance, or plasticity of the neuronal correlates of cognitive processing. TMS paradigms commonly used in cognitive studies can be grouped according to the type of TMS pulse used, mainly single-pulse TMS (SpTMS), paired-pulse TMS (PpTMS), and repetitive TMS (rTMS) (Fig. 2).

With SpTMS, a single magnetic pulse is delivered at a precise point in time during a cognitive task. Most TMS experiments using SpTMS have reported increases in reaction time (RT) usually longer than 50 ms more often than errors. This phenomenon has been described as if the brain has been "put on hold" for tens of milliseconds and, to a certain extent, creating a virtual lesion. In another words, SpTMS, under most conditions, only adds enough noise to delay the cognitive process of interest; but if the task is difficult enough and the magnetic stimulation intensity high enough, errors can occur. Because the SpTMS pulse acts

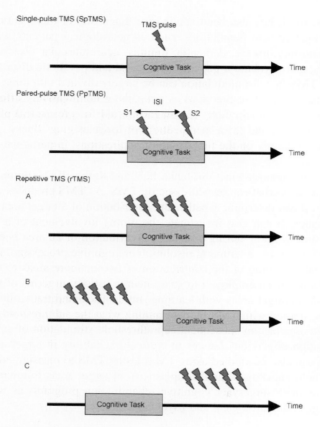

Fig. 2. Common TMS paradigms used in cognitive brain research. For single-pulse TMS (*SpTMS*) a TMS pulse is applied at a precise point in time during a cognitive task. Paired-pulse TMS (*PpTMS*) consists of two pulses separated in time by varying the interstimulus interval (ISI), where the first pulse (*S1*) is applied at subthreshold as the conditioning stimulus, and the second pulse (*S2*) is applied at suprathreshold as the test pulse. Repetitive TMS (*rTMS*) consists of trains of pulses that can be applied during (**A**), before (**B**), or after (**C**) a cognitive task

as a means of adding neural noise to a task, it is reasonable to assume that the size of the induced neural noise (i.e., the size of the effect) peaks at the onset of the pulse and diminishes throughout the duration of the effect. It has also been found that the critical time of applying the TMS pulse seems to be earlier than the time of critical differences in event-related potential (ERP) studies and closer to the latencies observed in single-unit studies (e.g., [8, 9]). Therefore, the neural disruption of the stage of processing may be earlier than that of the onset of critical operations due to the life-span of the neural disruption. However, it is important to note that it is still not clear how long the TMS pulse actually destabilizes neural processing. Therefore, caution should be taken regarding the correspondence between the effective TMS times and timing information from other methodologies [10].

PpTMS consists of two pulses separated in time by a varying interstimulus interval (ISI). This type of paradigm was originally designed to study motor functions [11], where the first stimulus pulse (S1) is applied below motor threshold as a conditioning stimulus. The second stimulus pulse (S2), known as the test pulse, is applied above motor threshold. In motor studies, it was found that a short ISI of 1–6 ms had an intracortical inhibition (ICI) effect, whereas a longer ISI (7–30 ms) showed intracortical facilitation [11–13]. Lowering the intensity of the test pulse also affected inhibition, and changing the coil orientation was found to affect excitation. However, the effective short and longer ISIs were found to be affected by drugs and neurological disease (e.g., [14]). Since then, PpTMS has been applied to psychological investigations [15]. For example, PpTMS paradigms have been applied to study mechanisms of disease causing changes in the excitability of cortical areas, such as in dementia [16] and multiple sclerosis [17]. Wassermann and colleagues [18], using the PpTMS paradigm, found that decreased ICI as measured by the conditioned motor evoked potential (MEP)/unconditioned MEP amplitude ratio was correlated with the domain of neuroticism in the NEO Personality Inventory Revised [19] in men with obsessive-compulsive disorder. In a motor imagery study, ICI was found to be decreased during motor imagery of dominant thumb abduction [20], evidence interpreted as supporting the use of motor imagery during rehabilitation of motor function. In a more recent study [21], applying a PpTMS paradigm to the right parietal cortex, visual awareness was modulated at a 3-s ISI with an increased number of failures to detect stimuli in the left visual field, whereas at a 5-s ISI the number of failures decreased when compared to SpTMS and sham stimulation.

An example of how the above TMS paradigms could be used to study chronometry in visual perception is well illustrated in the following series of studies. Amassian and colleagues [22] were the first to use SpTMS as a virtual lesion technique in the visual system. In this experiment, healthy subjects were required to identify the three letters of small and low-contrast trigrams (three-letter strings). SpTMS were applied using a round coil over the occipital cortex (approximately 2 cm above the inion) at various intervals after presentation of the trigram. Subjects showed impairment in letter recognition when the pulse was delivered at a delay (called stimulus-onset asynchrony, or SOA) of 60–140 ms. At SOAs of 80–100 ms, there was complete suppression of the visual perception of the trigrams (Fig. 3). This implied that only at these times after stimuli presentation was the occipital cortex making a critical contribution to letter recognition. Thus, SpTMS could provide insight as to when a given brain area is making a critical contribution to a targeted behavior. Later TMS studies of the visual system made a finer analysis of the contribution of the various visual areas to visual perception. Pascual-Leone and Walsh [23] used PpTMS to probe the timing interactions between V5 and V1. They paired V1–V5 and V5–V1 stimulation at various SOAs. Subjects reported perception of moving phosphene; but as the SOA of V5–V1 TMS was increased to 10–40 ms, perception of the phosphene was affected (Fig. 4). However, when they

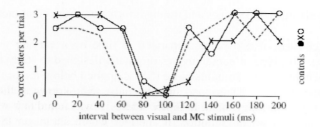

Fig. 3. TMS pulse to the occipital probe can suppress visual perception when it is applied 80–120 ms after stimulus presentation on three subjects using a circular coil. (From Amassian et al. [22], with permission)

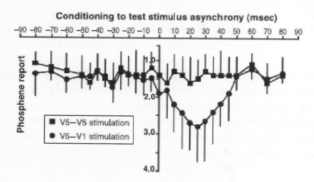

Fig. 4. Mean responses of all subjects ($n = 8$) to combined stimulation of visual cortical areas V5 and V1. The V5–V1 TMS asynchrony is displayed on the x-axis: Negative values indicate that V1 received TMS prior to V5, and positive values indicate that V1 was stimulated after V5. The subjects made four judgments: (1) elicited phosphene was present and moving; (2) phosphene was present, but subject not confident to judge whether it was moving; (3) phosphene was present but stationary; (4) no phosphene was observed. The perception of TMS-induced moving phosphene after stimulation of visual motion area V5 can be significantly suppressed by a second TMS pulse applied to the primary visual cortex (V1) 10–40 ms later. (From Pascual-Leone and Walsh [23], with permission)

paired V5–V5 stimulation at the same SOAs as above, there was no effect on the perception of phosphene. Thus, their finding showed that fast V5–V1 feedback projections are necessary for awareness of motion and established the time window of the back-projection from V5 to V1.

With rTMS, a repeating train of pulses are applied at a frequency >0.3 Hz. This is an arbitrary distinction based on the fact that early monophasic TMS machines needed about 3 s to charge up between successive pulses [24]. A train of repetitive pulses has been found to produce effects that can outlast the application of the stimulus for minutes and possibly hours. This characteristic is advantageous especially in cognitive studies where the display of a stimulus lasts ≥1 s, and it is necessary to disable neurons for longer than a few tens of milliseconds. Motor studies applying rTMS showed that the after-effects of rTMS depend on the frequency, intensity, and length of time for which the stimulation is given. In general, pulse frequencies ≤1 Hz have been termed low-frequency, or slow, rTMS; and rates >1 Hz are termed high-frequency, or fast, rTMS. There is evidence that low- and high-frequency stimulations could produce distinct effects on direct measures of brain activity and behavior. For example, using the resting motor threshold, rTMS, for 10 min at low frequency (1 Hz) have been found to decrease resting corticospinal excitability [25], whereas higher frequencies (>5–10 Hz) increased resting excitability [26, 27]. Although, the mechanism(s) underlying these after-effects of TMS are still unclear, it has been thought that the LTD for facilitation effects and LTP for inhibitory effects seem to be likely candidates for slow and fast rTMS, respectively. This suggestion is modeled after reduced animal models of cortical circuitry; however, there is a lack of direct evidence to either support or refute this claim [24].

rTMS techniques have been used as a "disruptor" of the normal pattern of cortical processing leading to impaired or delayed task performance in cognitive studies targeting various cortical areas (e.g., M1, frontal, parietal, temporal, and visual areas). The most widely used rTMS paradigm is to apply the trains of pulses while the subject performs a cognitive task of interest, known as online stimulation (Fig. 2A). This produces dramatic changes in behavior, such as speech arrest [28]; and it appears that the higher the rTMS frequency, the greater is the disruption of targeted brain region leading to greater behavioral effects. However, this approach suffers from potential nonspecific behavioral (e.g., muscle twitches) and attentional (e.g., distractions from the clicking noise from pulses) effects of concurrent stimulation that could adversely affect performance, rendering the results difficult to interpret [29]. Therefore, offline rTMS paradigms have been developed during the last few years to overcome some of these difficulties. One approach is to stimulate a site of interest for 5–10 min at 1 Hz prior to performing a cognitive task (Fig. 2B), thereby removing some of the nonspecific concurrent effects of TMS. This method has been applied to study visual imagery [30], visuospatial attention [31], visual working memory [32], motor learning [33], motor preparation [34], and grammatical processing [35]; it has also been proposed as a therapeutic technique in neurological and psychiatric patients [36]. More recently, a variation of this offline rTMS paradigm (Fig. 2C) has been applied to study cortical areas involved in consolidation of motor memories [37, 38].

The use of a combination of rTMS and SpTMS paradigms can help us first establish the site for stimulation and subsequently identify when this cortical site contributes to behavior. A good illustration has been provided by the TMS studies used to understand the role of the occipital cortex during Braille reading. Cohen and colleagues [39] found that short trains of rTMS to the occipital cortex induced errors in both identification of Braille and embossed Roman letters in early-onset blind subjects and distorted their tactile perception as well. In contrast,

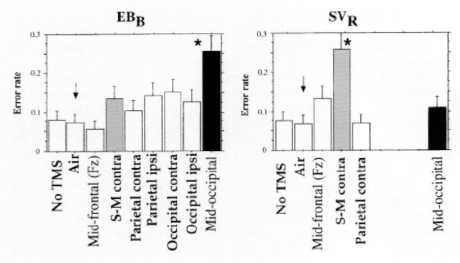

Fig. 5. Proportion of errors made in the discrimination of Braille characters during exposure to repetitive TMS. Short trains of repetitive TMS (10 Hz, 3 s) to occipital cortex disrupted tactile perception in early blind subjects (EB_B) reading Braille but not in sighted volunteers (SV_R) reading roman letters. *Open bars* indicate that stimulation at that position was not performed in that specific group. *Black bars* indicate error rates induced by stimulation of the mid-occipital position. *Gray bars* indicate error rates induced by stimulation of the contralateral sensorimotor cortex. *Asterisks* ($P < 0.001$) indicate scalp positions where significantly more errors occurred than in the controls *(arrows)*. *S-M*, sensorimotor cortex; *contra*, contralateral; *ipsi*, ipsilateral. (From Cohen et al. [39], with permission)

normal-sighted subjects showed no tactile deficits with occipital rTMS but had impaired visual performance (Fig. 5). These findings suggested that the visual cortex could be recruited for a role in somatosensory processing in people with blindness at an early age. The authors proposed that this early cross-modal plasticity may partly account for the superior tactile perceptual abilities in these subjects. In a separate study using SpTMS, Hamilton and Pascual-Leone [40] presented results from three blind Braille readers. They used a specially designed Braille simulator and presented real or nonsensical Braille stimuli to the pad of the subjects' reading index fingers. Following presentation of the Braille stimuli, they applied SpTMS to the appropriate sensorimotor cortex and striate occipital cortex at varying intervals (Fig. 6). They found that TMS to the sensorimotor cortex of blind subjects significantly reduces tactile detection when it is applied 20–40 ms after stimulus presentation, whereas TMS to their occipital cortex impairs tactile identification when it is applied 50–80 ms after stimulus presentation. The methods used in these studies show that rTMS can help us determine the areas making a critical contribution to an aspect of cognition, and SpTMS can help us pinpoint the time at which this contribution is made.

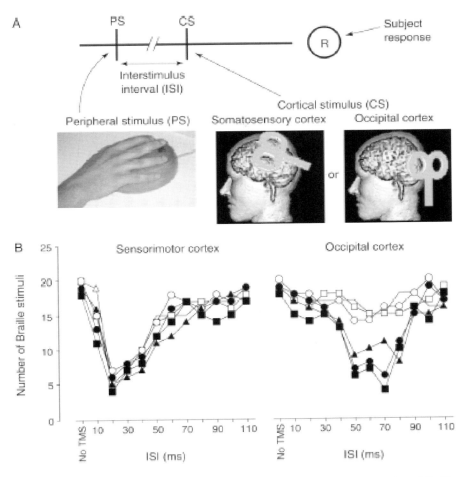

Fig. 6. Effects of cortical stimulation at various sites on detection and identification of Braille symbols. **A** Summary of the experimental design used. **B** Results for three subjects (*circles*, *triangles*, and *squares*, respectively). *Open symbols* show the number of stimuli detected by each subject in the two TMS conditions, regardless of whether real or nonsensical Braille stimuli were presented. *Filled symbols* show the number of correctly identified Braille stimuli ("real" or "nonsensical" and, if real, which Braille character) by each subject in the two TMS conditions, from among the stimuli that had been detected in the first place. (From Hamilton and Pascual-Leone [40], with permission)

Sham Stimulation

In all paradigms using TMS, there are several associated sensory sensations and sources of distraction that can potentially interfere with cognitive task demands. Therefore, to ascertain behavioral effects induced by TMS on the target brain area, we need to include control conditions, which is a common practice in psychological studies. These control conditions can be either stimulating a cortical site thought to be irrelevant to the task demands using sham stimulations (achieved by specially designed coils that produce similar physical effects without delivering any energy to the brain or orienting the coil perpendicular to the targeted area and minimizing cortical activation) or cleverly designed control task conditions that help rule out possible behavioral modulations due to irrelevant sensory effects (e.g., including a motor task condition for a study determining if the cerebellum is involved with higher cognitive function). For a more detailed discussion, see Jahanshahi and Rothwell [7].

Cognitive Enhancement

One application of TMS that has peaked the interest of neuroscientists is cognitive enhancement. This may sound paradoxical as TMS is known to act as a virtual lesion by disrupting organized neuronal firing. However, TMS can modulate the excitability of the targeted brain region beyond the duration of the stimulation train itself [41] (particularly rTMS when applied in short trains) and exert distant effects along functional neural networks [42]. It has been shown that 1-Hz stimulation for several minutes [26] or theta burst stimulation for a few tens of seconds [43] can alter brain excitability for a matter of minutes. One possible explanation for the apparent performance enhancement is that the shift in cortical excitability while interfering with local inhibitory circuits may allow either more or less access to brain networks relevant to the experimented task. Thus, functional gain can result from such modulation of activity in a distributed neuronal circuitry, improving performance in specific tasks.

Therefore, TMS has been targeted as a potential therapeutic modality for neurology, psychiatry, and rehabilitation as well as for enhancing performance in healthy subjects under certain circumstances. There are a growing number of studies showing TMS to have apparently improved performance in motor reaction times [44], mirror drawing [45], motor learning [46], story recall [44], verbal short-term memory tasks [29, 47], phonological memory [48], choice reaction task [49], reasoning [50], picture naming [51–53], mental rotation [54], and attention [55]. However, most of the beneficial effects have been short-lived, and a great deal of work remains to help us understand how TMS modulates neural mechanisms for a performance advantage.

Neuroscience Methods

During the last decade, we have seen the field of cognitive neuroscience develop at a feverish pace around the world. Neuroscience research institutes have been sprouting up in major universities and research centers at an unprecedented pace and urgency, emphasizing integrative research to unlock the secrets of the human brain. One main contributing force to this is the parallel advances in clinical neuroscience. The study of human patients with focal brain lesions provides another window on the role of particular brain regions in cognition and has been greatly facilitated by the advent of better structural neuroimaging, such as MRI, for lesion localization. The extension of studies to patients with more complex brain disorders using functional imaging promises to provide an approach to a systems-based understanding of human cognitive processes and their integration with emotional state. A better understanding of the molecular pathology that leads to clinical cognitive and behavioral disorders in humans can also prove relevant to understanding the function of the normal brain.

However, these fundamental advances would not have brought cognitive neuroscience to the present state without the advent of novel functional neuroimaging methods that enable scientists to measure regional brain activity online as normal humans perform cognitive tasks. These techniques include blood flow-based methods, such as fMRI and PET, and methods based on the electrical and magnetic fields associated with neural activity, such as event-related potentials (ERPs) and magnetoencephalography (MEG). Correlations between localized brain activity and behavior, however, do not of themselves establish that any brain area is essential for a particular task. The lesion technique can be regarded as the gold standard in establishing the need for a brain region in a cognitive process.

In neuropsychological studies, the ability to create a highly transient and reversible "virtual" lesion in a specific site of the brain makes TMS uniquely different from neuroimaging (MRI, fMRI, PET) and other correlation methods (ERP, EEG, MEG) in the realm of cognitive neuroscience research. Walsh and Cowey [56] provided an illustration of the "problem space" TMS occupies among popular neuroimaging techniques (Fig. 7). Behavioral tasks can evoke activation changes in a number of cortical and subcortical structures, but many of these structures are not necessary for task performance. A classic example is activation of the cingulate gyrus in a large number of neuroimaging experiments, including studies of language production. However, damage to this structure does not have the devastating consequences on speech that are caused by damage to Broca's area. TMS is also different from other brain mapping techniques in that it has great potential to become a less invasive, clinically therapeutic device. For example, in several studies, repetitive TMS has been applied to mental disorders, such as depression and schizophrenia, with efficacy similar to that of electroconvulsive therapy but without the need for anesthesia and without damaging effects on memory.

TMS and Brain Imaging

Apart from being a unique technique in itself, TMS is an ideal complement to functional neuroimaging. This is because it can test whether disruption of, for example fMRI-identified activations, produces a change in performance, thereby

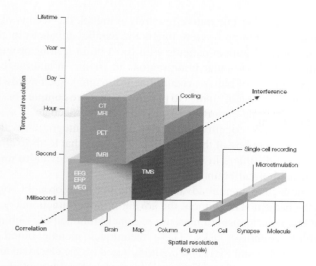

Fig. 7. Defining the problem space of TMS. The place of *TMS* in neuropsychological studies is best thought of as the "problem space" it occupies. This figure shows the spatial and temporal resolution of TMS compared with other neuroscience techniques. The ability of TMS to interfere with brain functions transiently and noninvasively in contrast to other correlative imaging techniques makes it unique. *CT*, computerized tomography; *EEG*, electroencephalography; *ERP*, event-related potential; *MEG*, magnetoencephalography; *fMRI*, functional magnetic resonance imaging; *PET*, positron emission tomography. (From Walsh and Cowey [56], with permission)

distinguishing which regions are necessary for task performance. As mentioned above, neuroimaging and patient studies are a similarly powerful combination of techniques; unlike patient studies, however, TMS can reveal the effects of disrupting a normally functioning neural circuit, whereas in patient studies brain reorganization and compensatory mechanisms may obscure the original contributions of the damaged structure. Furthermore, patient studies can reveal whether disruption of a structure affects performance, but they cannot reveal precisely when the structure makes its contribution during the task.

TMS can be applied before, during, and after functional neuroimaging (for a review see [57]). Functional imaging applied before TMS can help precisely localize the cortical area of activation during the performance of a certain task and help optimize the TMS coil position for the targeted area. This approach allows determination of the functional relevance of an activated cortical region for task execution by transiently disrupting and measuring resultant behavioral effects. For example, we used fMRI-guided TMS to determine if the right superior cerebellum contributes to verbal working memory [58]. This method enabled us to target subject-specific localizations of the right cerebellar activation during a verbal working memory task performed in the MRI scanner (Fig. 8). A frameless stereotactic navigation system was employed to interactively guide a 90-mm double-cone TMS coil that allowed deeper field penetration (3–4 cm in depth) [59–61] to the

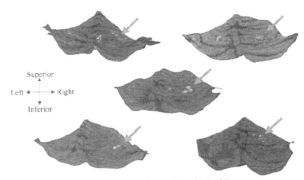

Fig. 8. Subject-specific localizations of right superior cerebellar activation obtained from a verbal working task. Surface renderings of the cerebellum used for TMS targeting lobule HVI/Crus I are depicted for five subjects. (From Desmond et al. [58], with permission)

desired region. The differences in the individual subject's activation in the right cerebellum seen on the functional maps suggest that activations within a region may be variable and using a pure coordinate system may not be adequate for spatial accuracy in this case [62].

When TMS is applied during the course of functional neuroimaging, we are able to evaluate the actual effects of TMS on cortical activity. This approach can help reveal in vivo information about regional cortical excitability [63, 64] and functional connectivity of the stimulated areas [42]. Synchronization of TMS and fMRI is of particular interest for cognitive neuroscience as it allows the possibility of stimulating neurocircuitry while monitoring changes in brain activity with high spatial and improved temporal resolution. However, technical problems present great challenges to such a combination, and only a handful of laboratories have managed such measurements without having to modify the TMS or MRI hardware [57]. Some of the technical challenges include the need for temporal synchronization of MRI data acquisition and TMS pulse application and the need to mount the TMS coil reliably within the MRI head coil. Another, more difficult challenge is to reduce the artifacts produced by the TMS pulse and the mere presence of the coil itself in the MRI scanner even when it is made of nonferromagnetic materials [65]. Artifacts generated by the TMS pulse can be avoided by interleaving the MRI acquisitions appropriately with the TMS pulse applications, such as requiring the MRI sequence to send a trigger signal to the TMS apparatus with every radiofrequency pulse excitation. Signal dropouts and geometric distortions of EPI images can also be markedly reduced when EPI is acquired in an orientation parallel to the coil plane [65].

Application of functional neuroimaging after TMS has been used to investigate the possible functional long-term effects of TMS on brain activation. For example, Lee and colleagues [66], using a water-based tracer ($H_2^{15}O$), found that regional cerebral blood flow (rCBF) was significantly increased locally at the stimulation site and extended to the nonstimulated hemisphere, mainly involving the premotor cortex during the resting state. In addition, the rCBF was found to be increased in the nonstimulated premotor cortex after 1-Hz rTMS. This approach allows us to gain empirical insights into functional cortical plasticity. Functional neuroimaging before and after TMS has also been employed to understand the cortical plasticity of the stimulated site. This technique is well illustrated by a study that investigated the effects of rTMS on memory performance in elders presenting with subjective memory complaints [67]. The authors administered two fMRI sessions with equivalent face–name memory tasks to 40 participants and applied a double-blind sham-controlled high-frequency rTMS paradigm to the prefrontal cortex between fMRI sessions. They found that only subjects who received active rTMS had significantly improved associative memory, which was accompanied by additional recruitment of right prefrontal and bilateral posterior cortical regions during the second fMRI session relative to the baseline scanning. However, it was not clear as to what specific prefrontal location the rTMS was stimulating to elicit the improved memory performance. Hence, together with an online navigator system and applying fMRI before and after TMS, we can further improve the precision of coil position and ascertain the stimulated location.

With the advances in MRI, we are now able to obtain tractography data of white matter fibers through the use of methods such as diffusion tensor imaging (DTI) and diffusion spectrum imaging (DSI). The combination of TMS with tractography data can help us further understand the correlation between structural and functional connectivity of various neuro-networks. Such techniques has been used to study the reorganization of motor function in those with congenital brain disorders [68] and to predict the functional potential in chronic stroke patients [69].

TMS and EEG

Neuronal connectivity defines the intricate patterns of cerebral activity associated with our sensory, motor, and cognitive functions. Whereas PET, fMRI, MEG, and EEG are able to provide maps of the distribution of activity in the brain, they are not particularly useful for determining corticocortical connections. TMS combined with high-resolution EEG enables noninvasive, direct study of functional connections between brain areas. By using high-resolution EEG to locate the changing pattern of the neuronal activity evoked by TMS, the initial cortical response reflecting cortical reactivity as well as the spread of activation from the stimulated site to other areas can be determined. EEG is a sensitive method for assessing cortical excitability induced by TMS, where brain responses were elicited even at subthreshold intensities of 60% of the individual's motor threshold (MT) [70, 71] compared to fMRI, with which changes in 80% of the MT has not been reliably detected [64, 72]. This ability to measure and describe TMS-induced cortical reactivity can help extend our understanding about the activation mechanisms of TMS. For example, Klimesch and colleagues [54, 73] applied EEG individual alpha frequency (IAF) as a "physiological template" to set the best tuning of rTMS for inducing transient improvement of cognitive performance. They found that when rTMS was delivered over the right parietal cortex at IAF − 1 Hz, also known as the individual's upper alpha frequency, it enhanced behavioral performance on a mental rotation task. They also found that during the actual performance of a task a large extent of event-related desynchronization (ERD) is correlated with good performance, and increased alpha reference power is associated with a large amount of ERD. Thus, the synchronization between rTMS and EEG rhythms appears to be an attractive technique for enhancing cognitive function, especially in the field of cognitive rehabilitation.

TMS-EEG has also been used to assess corticocortical connectivity (for a detailed review see [74]). TMS-evoked neuronal activity has been traced by EEG recorded with 60 scalp electrodes in six subjects [75]. The initial activity spread to surrounding cortical areas as well as to the contralateral cortex. In all six subjects, ipsilateral activity peaked over the precentral gyrus, supramarginal gyrus, or superior parietal lobule; activation over the contralateral cortex was also observed, appearing at 22 ± 2 ms (range 17–28 ms), with the maxima located at the precentral gyrus, superior frontal gyrus, and inferior parietal gyrus peaking at 24+ and −2 ms (range 18–29 ms) (Fig. 9). It has also been found that TMS-EEG connectivity patterns in healthy subjects are relatively repeatable and spatially specific [75], thus possibly providing a window to understand neuroplasticity in terms of

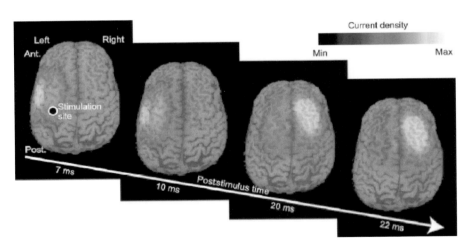

Fig. 9. Minimum–norm current-density estimates displayed above the cortical surface segmented from a three-dimensional magnetic resonance imaging set. It illustrates the spreading of neuronal activation after stimulation of the primary motor cortex for one subject. The *black circle* in the first image indicates the stimulation site. Activation spread from the stimulation site first to the neighboring cortical areas, mainly parietotemporally. Contralateral activation emerged at 20 ms and was strongest at 22 ms in frontal regions. The studied latency range was 7–30 ms after the stimulus. *Ant.*, anterior; *Post.*, posterior. (From Komssi et al. [74], with permission)

corticocortical connectivity in neurological diseases (e.g., brain injuries, degenerative disorders) and neuropharmacological interventions.

In recent years, TMS has also been combined with MEG for improved spatial resolution compared to that with EEG. This technique has been employed to address an important question in neuroplasticity during recovery of function: whether clinical recovery of sensory and motor functions is based on reestablishing previously damaged but not destroyed corticospinal connections or on reorganizing cortical somatotopy, where previously functionally silent or differently operating corticospinal fibers through the activation of newly established corticocortical connections are taking over the lost functions [76].

The combination of TMS-EEG is not without technical challenges, however, as the TMS pulse presents possible artifacts to be recorded by the scalp electrodes. Nevertheless, some of these challenges have been overcome; and currently such a commercial system has become available (eXimia EEG system, Nexstim, Helsinki, Finland).

Some Issues to Be Resolved

With the innovation of improved coil and stimulator designs, we have seen the applications of TMS flourish; correspondingly, the number of articles on TMS published in peer-reviewed journals has increased progressively during the last 20 years [77]. However, one must be careful when employing TMS in cognitive brain research as there are still a number of technical issues to be resolved (for a detailed review see [78]). For example, selecting the frequency, intensity, and duration of stimulation for a particular study is often difficult and arbitrary; and yet the success of a study is often based on these parameters. The most fundamental parameter in a TMS study is the intensity of stimulation for the subject; that is, at what intensity should we stimulate to interfere with a particular cognitive function at a specific site. This is a substantial problem, and there is no absolute method for determining this parameter.

The susceptibility of the brain area to stimulation can be influenced by the magnetic field strength and the excitability of the cortex. One could argue that because we know that the strength of the magnetic field produced by a coil decreases exponentially with distance we can use cortical depth (scalp-to-brain distance) of the site we want to stimulate to help us determine the intensity [79]. One needs to keep in mind that cortical excitability varies with the specific cortical site being stimulated and the behavioral task being performed. One popular method used is first to determine the lowest stimulation intensity capable of producing electrical activity in a muscle measured by electromyography (EMG) and obtain a "motor threshold" (MT) [80]. Subsequently, multiples of the MT can be used to normalize stimulation intensities across subjects. This standardized method allows comparison across experimental paradigms [81] if one is conducting motor studies because we can use muscle twitch to quantify excitability. However, generalizing this technique to other cortical regions could be erroneous, as the underlying assumption that the effects of TMS across cortical areas are correlated to one another appears to be incorrect. This is because phosphene thresholds (PTs) have been shown to be stable across times within individuals, but there appears to be no intraindividual correlation between PT and MT [81]. Therefore, it has been suggested that MT cannot be used to estimate the biological effects of TMS in cortical areas other than the motor cortex. This situation is further complicated by the fact that even within a single brain region the stimulation threshold for cortical disruption has been found to vary with the behavioral task (for an example see [82]).

Hence, other ways have been suggested to overcome this issue. Some studies have used a fixed intensity defined by the stimulator output; however, as with the MT technique, some subjects have cortical thresholds above the arbitrated fixed intensity that is capable of inducing a behavioral effect. Another method to determine stimulation intensity applied in the motor cortex is to obtain an input–output curve generated from measuring MEP amplitudes with incremental TMS intensity [83]. Assuming that nonmotor areas are also modulated by different TMS stimulation intensities evidenced by behavioral effects, we could apply a similar technique and apply parametric analysis to correlate the various stimulation intensities with behavioral responses. However, generating such an intensity tuning curve can be time-consuming when we need to perform it on each individual subject. In visual studies, some authors have titrated the intensity of TMS for a given experiment by using robust behavioral effects of stimulation on a separate task [84, 85]; however, TMS effects on one task may not correlate with its effects on another distinct task. Recently, TMS has been combined with other neuroimaging techniques, and the TMS effects on nonmotor areas can potentially be objectively quantified. Siebner and colleagues [86], using PET, showed that blood flow changes in the primary motor cortex are directly related to the intensity of stimulation. Using EEG monitoring, SpTMS applied over the motor cortex has been shown to be associated with one positive (P30) and two negative (N45, N100) scalp potentials [87]. In addition, it is found that the N45 component appears to be generated by the primary motor cortex, and its amplitude is correlated with the intensity of the TMS pulse. Hence, it is possible to use TMS-induced blood flow changes and scalp potentials as surrogate markers of the TMS effects on nonmotor cortical areas and functionally determine the intensity of stimulation. However, it is still not clear what the relation is between a cognitive task (and its potential disruption) and changes in cerebral blood flow, oxygenation, or scalp potentials.

Another substantial problem using TMS in cognitive studies is anatomical localization, or determining where the stimulation is being applied. This is because the relation between scalp position and a given brain area is variable across individuals. One method often used in motor and visual studies is to define the coil position based on objective output measures, such as muscle twitch or induction of phosphenes. However, complications of the coil placement arise when no overt responses are elicited from the cortical area of interest. A possible solution is to locate brain areas relative to those that have a reasonable certain position. For example, the dorsal lateral prefrontal cortex (DLPFC) has been defined as 5 cm anterior to the thumb representation over the primary MT using the Talairach atlas [88, 89]. However, validation studies comparing the Talairach position against the known brain anatomy of individual MRI scans found that the final coil location relative to the underlying Brodmann's area is quite variable [90]; Herwig and colleagues [91] showed only 7 of 22 subjects had accurate coil positions using this method. Frameless stereotactic systems adapted from neurosurgery have been employed to provide online information about coil location by first obtaining an MRI scan of the subject [92–94]. This image-guided TMS technique allows constant visualization of the coil placement in relation to the subject's brain. However, this assumes that gross anatomical features are related to functional subdivisions that may be problematic. More recently, the use of frameless stereotaxy with functional neuroimaging (e.g., fMRI) allows activated sites to be targeted for stimulation and reduces the influence of interindividual anatomical variability.

Lastly, one has to be careful when interpreting behavioral results in relation to the site of stimulation. This is because each brain area is coupled through anatomical connections and projections with a vast number of other areas. Therefore, stimulating one area of the brain may have functional consequences not only at that site but throughout the neural circuitry, especially with the use of rTMS [95, 96]. This is where the integration of tractography data with TMS may provide further understanding of this problem. The above are just some highlights of the challenges we face with the current technology in TMS. It is therefore critical to have hypothesis-driven and carefully designed experiments that acknowledge the current limitations of TMS.

Safety and Persistence of TMS

The main concern about the safety of TMS is the fear of producing short- or long-term damage to the brain by kindling effects, thereby inducing seizures. Therefore, TMS as an experimental technique has been monitored closely for safety issues. Now, after 20 years, tens of thousands of subjects have undergone the procedure for diagnostic and investigative purposes, as well as therapeutic trials, without major adverse reactions or side effects. SpTMS have been shown to be extremely safe and has been labeled a procedure of nonsignificant risk by the Food and Drug Administration (FDA) in the United States. rTMS has been reported to induce seizures in a small percentage of healthy subjects [29, 97, 98] and potentially more pronounced and longer-term effects on neural functioning. However, since the publication of the safety guidelines for rTMS [29] and apparent adherence to these guidelines by the TMS community, there have been no seizures reported in studies conducted on healthy subjects.

A recent review of the literature for adverse effects of rTMS on nonmotor cortical areas in healthy and patient participants revealed that headache was the most common complaint and is more frequent with frontal rTMS [99]; however, it was tolerable. Other adverse effects consisted of nausea, neck pain, and peripheral muscle twitches. The authors reported that serious adverse side effects were rare and consisted of two seizures and four instances of psychotic symptoms induced by rTMS over the DLPFC of patients with depression. It was concluded that, overall, rTMS to nonmotor areas appears to be safe with few, generally mild adverse effects.

Nevertheless, the mention of major safety guidelines when conducting TMS experiments seems warranted. All participants should be screened for TMS contraindications, such as a personal or family history of seizures, brain lesions, or other neurological conditions that are not part of the investigation. The use of ear plugs is recommended during all experiments, as the TMS pulse generates a loud noise, and temporary elevations of auditory thresholds have been reported [44]. Persistence of the stimulation effect from rTMS has not been well studied and may pose an ethical issue. We want the therapeutic and enhancement effects of TMS to last beyond the TMS session, but there is no discounting that deleterious effects may linger as well. Therefore, it is recommended that an appropriate neuropsychological or neurological battery of tests be administered before and several times after rTMS [29].

Conclusion

Transcranial magnetic stimulation has proved to be a valuable tool for investigating normal human neurophysiology and holds promise for diagnostic and therapeutic purposes. It has provided us a possible way to manipulate the brain, rather than passively viewing what it does (such as with functional neuroimaging). The complementary use of TMS with other brain mapping techniques will allow potential breakthroughs in the study of functional organization of the human brain. Hence, TMS as a brain mapping technique is undoubtedly gaining ground as an important tool for exploring cognitive neuroscience and is being established as an indispensable technique in neuroscience research.

References

1. Barker AT, Jalinous R, Freeston IL(1985) Non-invasive magnetic stimulation of human motor cortex. Lancet 1:1106–1107.
2. Cowey A (2005) The Ferrier Lecture 2004: what can transcranial magnetic stimulation tell us about how the brain works? Philos Transact R Soc Lond 360:1185–1205.
3. Jalinous R (2002) Principles of magnetic stimulator design. In: Pascual-Leone A, Davey NJ, Rothwell J, et al (eds) Handbook of transcranial magnetic stimulation. Oxford University Press, New York, pp 30–38.
4. Cowey A, Walsh V (2000) Magnetically induced phosphenes in sighted, blind and blindsighted observers. Neuroreport 11:3269–3273.

5. Barker AT (1999) The history and basic principles of magnetic nerve stimulation. Electroencephalogr Clin Neurophysiol Suppl 51:3–21.
6. Illes J, Gallo M, Kirschen MP (2006) An ethics perspective on transcranial magnetic stimulation (TMS) and human neuromodulation. Behav Neurol 17:149–157.
7. Jahanshahi M, Rothwell J (2000) Transcranial magnetic stimulation studies of cognition: an emerging field. Exp Brain Res 131:1–9.
8. Ashbridge E, Walsh V, Cowey A (1997) Temporal aspects of visual search studied by transcranial magnetic stimulation. Neuropsychologia 35:1121–1131.
9. O'Shea J, Muggleton NG, Cowey A, et al (2004) Timing of target discrimination in human frontal eye fields. J Cogn Neurosci 16:1060–1067.
10. Walsh V, Pascual-Leone A (2003) Transcranial magnetic stimulation: a neurochronometrics of mind. MIT Press, Camridge, MA, pp 65–94.
11. Kujirai T, Caramia MD, Rothwell JC, et al (1993) Corticocortical inhibition in human motor cortex. J Physiol 471:501–519.
12. Vucic S, Howells J, Trevillion L, et al (2006) Assessment of cortical excitability using threshold tracking techniques. Muscle Nerve 33:477–486.
13. Maeda F, Gangitano M, Thall M, et al (2002) Inter- and intra-individual variability of paired-pulse curves with transcranial magnetic stimulation (TMS). Clin Neurophysiol 113: 376–382.
14. Lefaucheur JP (2005) Motor cortex dysfunction revealed by cortical excitability studies in Parkinson's disease: influence of antiparkinsonian treatment and cortical stimulation. Clin Neurophysiol 116:244–253.
15. Oliveri M, Caltagirone C, Filippi MM, et al (2000) Paired transcranial magnetic stimulation protocols reveal a pattern of inhibition and facilitation in the human parietal cortex. J Physiol 529(Pt 2):461–468.
16. Pierantozzi M, Panella M, Palmieri MG, et al (2004) Different TMS patterns of intracortical inhibition in early onset Alzheimer dementia and frontotemporal dementia. Clin Neurophysiol 115:2410–2418.
17. Mainero C, Inghilleri M, Pantano P, et al (2004) Enhanced brain motor activity in patients with MS after a single dose of 3,4-diaminopyridine. Neurology 62:2044–2050.
18. Wassermann EM, Greenberg BD, Nguyen MB, et al (2001) Motor cortex excitability correlates with an anxiety-related personality trait. Biol Psychiatry 50:377–382.
19. Costa PT, McCrae RR (1992) Revised NEO Personality Inventory (NEO-PI-R) and NEO Five-Factor Inventory (NEO-FFI) Professional Manual. Psychological Assessment Resources, Odessa, FL.
20. Stinear CM, Byblow WD (2003) Motor imagery of phasic thumb abduction temporally and spatially modulates corticospinal excitability. Clin Neurophysiol 114:909–914.
21. Koch G, Oliveri M, Torriero S, et al (2005) Modulation of excitatory and inhibitory circuits for visual awareness in the human right parietal cortex. Exp Brain Res 160:510–516.
22. Amassian VE, Cracco RQ, Maccabee PJ, et al (1989) Suppression of visual perception by magnetic coil stimulation of human occipital cortex. Electroencephalogr Clin Neurophysiol 74:458–462.
23. Pascual-Leone A, Walsh V (2001) Fast backprojections from the motion to the primary visual area necessary for visual awareness. Science 292:510–512.
24. Rothwell J (2003) Techniques of transcranial magnetic stimulation. In: Boniface S, Ziemann U (eds) Plasticity in the human nervous system: investigations with transcranial magnetic stimulation. Cambridge University Press, Cambridge, UK, pp 26–61.
25. Touge T, Gerschlager W, Brown P, et al (2001) Are the after-effects of low-frequency rTMS on motor cortex excitability due to changes in the efficacy of cortical synapses? Clin Neurophysiol 112:2138–2145.
26. Chen R, Classen J, Gerloff C, et al (1997) Depression of motor cortex excitability by low-frequency transcranial magnetic stimulation. Neurology 48:1398–1403.
27. Maeda F, Keenan JP, Tormos JM, et al (2000) Modulation of corticospinal excitability by repetitive transcranial magnetic stimulation. Clin Neurophysiol 111:800–805.

28. Pascual-Leone A, Gates J, Dhuna A (1991) Induction of speech arrest and counting errors with rapid-rate transcranial magnetic stimulation. Neurology 41:697–702.
29. Wassermann EM (1998) Risk and safety of repetitive transcranial magnetic stimulation: report and suggested guidelines from the International Workshop on the Safety of Repetitive Transcranial Magnetic Stimulation, June 5–7, 1996. Electroencephalogr Clin Neurophysiol 108:1–16.
30. Kosslyn SM, Pascual-Leone A, Felician O, et al (1999) The role of area 17 in visual imagery: convergent evidence from PET and rTMS. Science 284:167–170.
31. Kim Y-H, Min S-J, Ko M-H, et al (2005) Facilitating visuospatial attention for the contralateral hemifield by repetitive TMS on the posterior parietal cortex. Neurosci Lett 382:280–285.
32. Hilgetag CC, Theoret H, Pascual-Leone A (2001) Enhanced visual spatial attention ipsilateral to rTMS-induced 'virtual lesions' of human parietal cortex. Nat Neurosci 4:953–957.
33. Robertson EM, Tormos JM, Maeda F, et al (2001) The role of the dorsolateral prefrontal cortex during sequence learning is specific for spatial information. Cereb Cortex 11:628–635.
34. Rounis E, Yarrow K, Rothwell JC (2007) Effects of rTMS conditioning over the fronto-parietal network on motor versus visual attention. J Cogn Neurosci 19:513–524.
35. Shapiro KA, Pascual-Leone A, Mottaghy FM, et al (2001) Grammatical distinctions in the left frontal cortex. J Cogn Neurosci 13:713–720.
36. Terao Y, Ugawa Y (2006) Studying higher cerebral functions by transcranial magnetic stimulation. Clin Neurophysiol 59(suppl):9–17.
37. Muellbacher W, Ziemann U, Wissel J, et al (2002) Early consolidation in human primary motor cortex. Nature 415:640–644.
38. Robertson EM, Press DZ, Pascual-Leone A (2005) Off-line learning and the primary motor cortex. J Neurosci 25:6372–6378.
39. Cohen LG, Celnik P, Pascual-Leone A, et al (1997) Functional relevance of cross-modal plasticity in blind humans. Nature 389:180–183.
40. Hamilton RH, Pascual-Leone A (1998) Cortical plasticity associated with Braille learning. Trends Cogn Sci 2:168–174.
41. Pascual-Leone A, Cohen LG, Brasil-Neto JP, et al (1994) Differentiation of sensorimotor neuronal structures responsible for induction of motor evoked potentials, attenuation in detection of somatosensory stimuli, and induction of sensation of movement by mapping of optimal current directions. Electroencephalogr Clin Neurophysiol 93:230–236.
42. Paus T, Jech R, Thompson CJ, et al (1997) Transcranial magnetic stimulation during positron emission tomography: a new method for studying connectivity of the human cerebral cortex. J Neurosci 17:3178–3184.
43. Huang YZ, Edwards MJ, Rounis E, et al (2005) Theta burst stimulation of the human motor cortex. Neuron 45:201–206.
44. Pascual-Leone A, Houser CM, Reese K, et al (1993) Safety of rapid-rate transcranial magnetic stimulation in normal volunteers. Electroencephalogr Clin Neurophysiol 89:120–130.
45. Balslev D, Christensen LO, Lee JH, et al (2004) Enhanced accuracy in novel mirror drawing after repetitive transcranial magnetic stimulation-induced proprioceptive deafferentation. J Neurosci 24:9698–9702.
46. Kim YH, Park JW, Ko MH, et al (2004) Facilitative effect of high frequency subthreshold repetitive transcranial magnetic stimulation on complex sequential motor learning in humans. Neurosci Lett 367:181–185.
47. Kohler S, Paus T, Buckner RL, et al (2004) Effects of left inferior prefrontal stimulation on episodic memory formation: a two-stage fMRI-rTMS study. J Cogn Neurosci 16:178–188.
48. Kirschen MP, Davis-Ratner MS, Jerde TE, et al (2006) Enhancement of phonological memory following transcranial magnetic stimulation (TMS). Behav Neurol 17:187–194.
49. Evers S, Bockermann I, Nyhuis PW (2001) The impact of transcranial magnetic stimulation on cognitive processing: an event-related potential study. Neuroreport 12:2915–2918.

50. Boroojerdi B, Phipps M, Kopylev L, et al (2001) Enhancing analogic reasoning with rTMS over the left prefrontal cortex. Neurology 56:526–528.
51. Mottaghy FM, Hungs M, Brugmann M, et al (1999) Facilitation of picture naming after repetitive transcranial magnetic stimulation. Neurology 53:1806–1812.
52. Topper R, Mottaghy FM, Brugmann M, et al (1998) Facilitation of picture naming by focal transcranial magnetic stimulation of Wernicke's area. Exp Brain Res 121:371–378.
53. Mottaghy FM, Sparing R, Topper R (2006) Enhancing picture naming with transcranial magnetic stimulation. Behav Neurol 17:177–186.
54. Klimesch W, Sauseng P, Gerloff C (2003) Enhancing cognitive performance with repetitive transcranial magnetic stimulation at human individual alpha frequency. Eur J Neurosci 17:1129–1133.
55. Fecteau S, Pascual-Leone A, Theoret H (2006) Paradoxical facilitation of attention in healthy humans. Behav Neurol 17:159–162.
56. Walsh V, Cowey A (2000) Transcranial magnetic stimulation and cognitive neuroscience. Nat Rev Neurosci 1:73–79.
57. Sack AT, Linden DE (2003) Combining transcranial magnetic stimulation and functional imaging in cognitive brain research: possibilities and limitations. Brain Res Brain Res Rev 43:41–56.
58. Desmond JE, Chen SH, Shieh PB (2005) Cerebellar transcranial magnetic stimulation impairs verbal working memory. Ann Neurol 58:553–560.
59. Maccabee PJ, Eberle L, Amassian VE, et al (1990) Spatial distribution of the electric field induced in volume by round and figure "8" magnetic coils: relevance to activation of sensory nerve fibers. Electroencephalogr Clin Neurophysiol 76:131–141.
60. Terao Y, Ugawa Y, Hanajima R, et al (2000) Predominant activation of I1-waves from the leg motor area by transcranial magnetic stimulation. Brain Res 859:137–146.
61. Terao Y, Ugawa Y, Sakai K, et al (1994) Transcranial stimulation of the leg area of the motor cortex in humans. Acta Neurol Scand 89:378–383.
62. Sparing R, Buelte D, Meister IG, et al (2007) Transcranial magnetic stimulation and the challenge of coil placement: a comparison of conventional and stereotaxic neuronavigational strategies. Hum Brain Mapp [Epub ahead of print]
63. Bohning DE, Pecheny AP, Epstein CM, et al (1997) Mapping transcranial magnetic stimulation (TMS) fields in vivo with MRI. Neuroreport 8:2535–2538.
64. Bohning DE, Shastri A, McConnell KA, et al (1999) A combined TMS/fMRI study of intensity-dependent TMS over motor cortex. Biol Psychiatry 45:385–394.
65. Baudewig J, Paulus W, Frahm J (2000) Artifacts caused by transcranial magnetic stimulation coils and EEG electrodes in T(2)*-weighted echo-planar imaging. Magn Reson Imaging 18:479–484.
66. Lee L, Siebner HR, Rowe JB, et al (2003) Acute remapping within the motor system induced by low-frequency repetitive transcranial magnetic stimulation. J Neurosci 23:5308–5318.
67. Sole-Padulles C, Bartres-Faz D, Junque C, et al (2006) Repetitive transcranial magnetic stimulation effects on brain function and cognition among elders with memory dysfunction: a randomized sham-controlled study. Cereb Cortex 16:1487–1493.
68. Kim YH, Jang SH, Han BS, et al (2004) Ipsilateral motor pathway confirmed by diffusion tensor tractography in a patient with schizencephaly. Neuroreport 15:1899–1902.
69. Stinear CM, Barber PA, Smale PR, et al (2007) Functional potential in chronic stroke patients depends on corticospinal tract integrity. Brain 130:170–180.
70. Kahkonen S, Komssi S, Wilenius J, et al (2005) Prefrontal transcranial magnetic stimulation produces intensity-dependent EEG responses in humans. Neuroimage 24:955–960.
71. Komssi S, Kahkonen S, Ilmoniemi RJ (2004) The effect of stimulus intensity on brain responses evoked by transcranial magnetic stimulation. Hum Brain Mapp 21:154–164.
72. Nahas Z, Lomarev M, Roberts DR, et al (2001) Unilateral left prefrontal transcranial magnetic stimulation (TMS) produces intensity-dependent bilateral effects as measured by interleaved BOLD fMRI. Biol Psychiatry 50:712–720.

73. Klimesch W, Doppelmayr M, Hanslmayr S (2006) Upper alpha ERD and absolute power: their meaning for memory performance. Prog Brain Res 159:151–165.
74. Komssi S, Kahkonen S (2006) The novelty value of the combined use of electroencephalography and transcranial magnetic stimulation for neuroscience research. Brain Res Rev 52:183–192.
75. Komssi S, Aronen HJ, Huttunen J, et al (2002) Ipsi- and contralateral EEG reactions to transcranial magnetic stimulation. Clin Neurophysiol 113:175–184.
76. Rossini PM, Pauri F (2000) Neuromagnetic integrated methods tracking human brain mechanisms of sensorimotor areas 'plastic' reorganisation. Brain Res Brain Res Rev 33:131–154.
77. Rossini PM, Rossi S (2007) Transcranial magnetic stimulation: diagnostic, therapeutic, and research potential. Neurology 68:484–488.
78. Robertson EM, Theoret H, Pascual-Leone A (2003) Studies in cognition: the problems solved and created by transcranial magnetic stimulation. J Cogn Neurosci 15:948–960.
79. McConnell KA, Nahas Z, Shastri A, et al (2001) The transcranial magnetic stimulation motor threshold depends on the distance from coil to underlying cortex: a replication in healthy adults comparing two methods of assessing the distance to cortex. Biol Psychiatry 49: 454–459.
80. Rossini PM, Barker AT, Berardelli A, et al (1994) Non-invasive electrical and magnetic stimulation of the brain, spinal cord and roots: basic principles and procedures for routine clinical application: report of an IFCN committee. Electroencephalogr ClinNeurophysiol 91: 79–92.
81. Stewart LM, Walsh V, Rothwell JC (2001) Motor and phosphene thresholds: a transcranial magnetic stimulation correlation study. Neuropsychologia 39:415–419.
82. Theoret H, Kobayashi M, Ganis G, et al (2002) Repetitive transcranial magnetic stimulation of human area MT/V5 disrupts perception and storage of the motion aftereffect. Neuropsychologia 40:2280–2287.
83. Chen R (2000) Studies of human motor physiology with transcranial magnetic stimulation. Muscle Nerve 9:S26–S32.
84. Mulleners WM, Chronicle EP, Palmer JE, et al (2001) Suppression of perception in migraine: evidence for reduced inhibition in the visual cortex. Neurology 56:178–183.
85. Rushworth MF, Ellison A, Walsh V (2001) Complementary localization and lateralization of orienting and motor attention. Nat Neurosci 4:656–661.
86. Siebner HR, Takano B, Peinemann A, et al (2001) Continuous transcranial magnetic stimulation during positron emission tomography: a suitable tool for imaging regional excitability of the human cortex. Neuroimage 14:883–890.
87. Paus T, Sipila PK, Strafella AP (2001) Synchronization of neuronal activity in the human primary motor cortex by transcranial magnetic stimulation: an EEG study. J Neurophysiol 86:1983–1990.
88. Pascual-Leone A, Wassermann EM, Grafman J, et al (1996) The role of the dorsolateral prefrontal cortex in implicit procedural learning: experimental brain research Exp Hirnforsch 107:479–485.
89. Pascual-Leone A, Rubio B, Pallardo F, et al (1996) Rapid-rate transcranial magnetic stimulation of left dorsolateral prefrontal cortex in drug-resistant depression. Lancet 348:233–237.
90. Pascual-Leone A, Bartres-Faz D, Keenan JP (1999) Transcranial magnetic stimulation: studying the brain-behaviour relationship by induction of 'virtual lesions'. Philos Transact R Soc Lond 354:1229–1238.
91. Herwig U, Padberg F, Unger J, et al (2001) Transcranial magnetic stimulation in therapy studies: examination of the reliability of "standard" coil positioning by neuronavigation. Biol Psychiatry 50:58–61.
92. Gugino LD, Romero JR, Aglio L, et al (2001) Transcranial magnetic stimulation coregistered with MRI: a comparison of a guided versus blind stimulation technique and its effect on evoked compound muscle action potentials. Clin Neurophysiol 112:1781–1792.
93. Herwig U, Schonfeldt-Lecuona C, Wunderlich AP, et al (2001) The navigation of transcranial magnetic stimulation. Psychiatry Res 108:123–131.

94. Paus T (1999) Imaging the brain before, during, and after transcranial magnetic stimulation. Neuropsychologia 37:219–224.
95. Munchau A, Bloem BR, Irlbacher K, et al (2002) Functional connectivity of human premotor and motor cortex explored with repetitive transcranial magnetic stimulation. J Neurosci 22:554–561.
96. Gerschlager W, Siebner HR, Rothwell JC (2001) Decreased corticospinal excitability after subthreshold 1 Hz rTMS over lateral premotor cortex. Neurology 57:449–455.
97. Chen R, Gerloff C, Classen J, et al (1997) Safety of different inter-train intervals for repetitive transcranial magnetic stimulation and recommendations for safe ranges of stimulation parameters. Electroencephalogr Clin Neurophysiol 105:415–421.
98. Green RM, Pascual-Leone A, Wasserman EM (1997) Ethical guidelines for rTMS research. IRB 19:1–7.
99. Machii K, Cohen D, Ramos-Estebanez C, et al (2006) Safety of rTMS to non-motor cortical areas in healthy participants and patients. Clin Neurophysiol 117:455–471.

Spectral Analysis of fMRI Signal and Noise

Chien-Chung Chen[1] and Christopher W. Tyler[2]

Summary

We analyzed the noise in functional magnetic resonance imaging (fMRI) scans of the human brain during rest. The noise spectrum in the cortex is well fitted by a model consisting of two additive components: flat-spectrum noise that is uniform throughout the MRI image and frequency-dependent biological noise that is localized to the neural tissue and declines from low to high temporal frequencies. We show that the frequency-dependent component is well fitted by the f^{-p} model with $0 < p < 1$ throughout the measured frequency range. The parameters of the model indicate that the characteristic noise is not attributable to the temporal filtering of the hemodynamic response but is an inherent property of the blood oxygenation level-dependent (BOLD) signal. We then analyzed the power spectrum of the BOLD signal for various cognitive tasks. The signal-to-noise ratio of a typical fMRI experiment peaks at around 0.04 Hz.

Key words Blood oxygenation level-dependent (BOLD), Rest scan, Cortex, Stochastic process

Introduction

In a typical functional magnetic resonance imaging (fMRI) experiment, the local change in blood oxygenation level-dependent (BOLD) activity in the brain of a participant over time was measured while the participant engaged in certain sensorimotor or cognitive tasks [1]. Using this technique, the relative activation in cortical regions on a spatial scale of millimeters of cortex can be compared for a wide variety of tasks designed to isolate aspects of neural function—from early sensory processing to complex decision making and action planning. One major limiting factor for fMRI is the low signal-to-noise ratio (SNR) of the BOLD

---

[1]Department of Psychology, National Taiwan University, Taipei 106, Taiwan
[2]The Smith-Kettlewell Eye Research Institute, San Francisco, CA 94115, USA

activity. Here, "signal" means the BOLD activity driven by the psychological tasks designated by the experimenters, and "noise" means the activity unrelated to such tasks. In a typical fMRI experiment, hemodynamic modulation induced by a psychological task is about 1%–5% around the mean activation level. The random variation of the BOLD activation can easily reach 2%–4% in magnitude [2]. Thus, fMRI data analysis is highly susceptible to noise. The purpose of this study was to analyze both the signal and noise power spectra of BOLD activation during a typical fMRI experiment time course. This knowledge has implications for better fMRI experimental designs and data analysis.

Because the noise is defined as the BOLD activity in the brain unrelated to any specific psychological task designated by the experimenter, the noise spectrum can be measured with a rest scan—that is, when participants are not required to perform any task except close their eyes and remain still throughout a scan run. The Fourier transform of such rest scan time series measures the average power spectrum of BOLD noise. It is important to understand the properties of the noise spectra to optimize the design of the activation signals for a maximal SNR.

It has been shown that the power spectrum of BOLD noise in the brain decreases with temporal frequency [3–6]. In particular, Zarahn et al. [4] fitted the power spectrum with arbitrary expressions containing a negative slope of −1 and called it 1/f noise. Here we offer a more detailed analysis of the noise properties in various brain structures. First, one might expect that there would be an irreducible component of equipment noise in the magnetic resonance detection process, which is expected by default to be Gaussian white noise by virtue of its derivation from multiples sources in the measurement system. This white noise must be combined with a characteristic hemodynamic noise component with a negative slope set by the properties of the oxygen resonance in the magnetic field. It is not clear what aspect of the biological signal generates this negative slope.

We therefore analyzed the temporal frequency spectrum of the BOLD signal during rest scans in two ways. To isolate the nonbiological component, we characterized both the extracephalic fMRI spectrum for voxels selected from the space of the fMRI reconstruction lying outside the head and the spectrum for a water phantom. In the head, we segmented the voxels into those lying entirely within the cortical gray matter, those entirely within the white matter, those outside the brain, and a residual set on the basis of the mean BOLD level. As the cortical hemodynamic noise is most relevant for fMRI recording, we then characterized the (cortical) fMRI spectrum for gray matter voxels. For comparison, we also determined the fMRI spectrum for white matter voxels and extracephalic voxels.

The second approach was to analyze the slope of the noise power spectrum reflecting the stochastic process of the noise-generating mechanisms. At the two extremes, a flat spectrum of slope zero ("white noise") is a characteristic of a memoryless stochastic process, and a slope of −1 is a result of a long-memory random walk process. The slope of the noise power spectrum thus reveals much information on the mechanisms of the noise that fMRI measures. We therefore fitted a multicomponent noise model to the data from rest scans.

The signal dynamics can be measured in both the frequency domain and the time domain. Bandettini [7] instructed his participants to move their fingers for a few seconds (on-epoch) and then rest for the same amount of time (off-epoch); they were then told to repeat the on–off switch several times. He showed that the BOLD activation change in the motor cortex between the on- and off-epochs increased with the period of on–off cycles. Boynton et al. [8] showed similar frequency-dependent activation change in the visual cortex to visual stimuli. In general, these power spectra showed a low pass characteristic.

In the time domain, the BOLD activation change following a brief psychological event is well known. The typical BOLD activation in the responsive cortical area shows an increase shortly after the stimulus onset and a post-stimulation undershoot after the stimulus offset [9, 10]. The duration of the activation increment depends on the stimulus duration. However, because the hemodynamic response is essentially linear under most circumstances [8], one can derive an empirical hemodynamic response function by deconvolving the activation time series and the sequence of stimulation [10] and, in turn, the power spectrum of the hemodynamic signal by its Fourier transform. Here we use the differential power spectrum of the fMRI signal to that of the noise to determine the temporal frequency of response with the best SNR. This result should help indicate the best time course for the fMRI experiments.

Noise

Spectra for Hardware Noise

We first investigated noise from nonbiological sources. Figure 1 shows the power spectrum of extracephalic voxels during rest scans. The time series contained 72

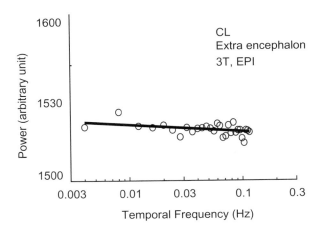

Fig. 1. Extracephalic resting spectra. Data show temporal frequency spectrum from voxels in the extracephalic space during the rest scan of observer C.L. Note that the functions are close to flat. *EPI*, echo planar imaging

volumes acquired with 3-s TR. The extracephalic voxels in the analysis were selected from a contiguous zone along the edges of the slice images. Those voxels were at least 30 mm away from the closest encephalic voxels in the same-slice image. The time series data were acquired during a rest scan in eight participants. For images acquired with an echo planar imaging (EPI) sequence [11], we avoided choosing extracephalic voxels in the phase selection direction of the images to minimize the influence of ghost images. The mean intensity of extracephalic voxels was less than 10% of the mean intensity of the gray matter voxels (i.e., close to black in the images).

Figure 1 shows an example of the extraencephalic spectrum. The amplitude spectrum outside the brain is essentially flat up to the Nyquist frequency of 0.17 Hz. For the fit of a straight line over the full spectrum in this log–log plot, the average slope for our eight participants was −0.04 ± 0.08, showing no significant deviation from a white noise spectrum.

There is concern that the noise amplitude in a voxel increases with the mean intensity of that voxel. In this case, because the cephalic voxels have much higher intensity than the extracephalic voxels, the power spectra from extracephalic voxels may not reflect the full contribution of the hardware noise. Hence, we acquired data on water phantoms with the same time course as in the rest scan.

Although the mean intensity of the voxels in a water phantom was about 40 times the mean intensity of the extracephalic voxels in our scans, their power spectra showed similar behavior (Fig. 2). Across the spectrum, the average slope was 0.0007 ± 0.0027 for the two water phantom scans. This white noise behavior is expected from the prediction based on thermal and other noise sources in the scanner.

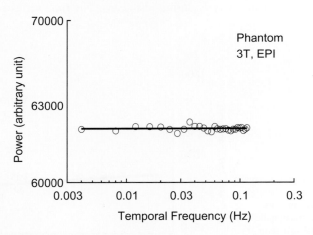

Fig. 2. Extracephalic resting spectrum. Data show temporal frequency spectrum from voxels in a water phantom. Note that the function is close to flat

Noise Spectra in the Brain

Figure 3 shows the power spectrum of time series data acquired in rest scans of eight participants. Six participants were scanned in the National Taiwan University magnetic resonance imaging/magnetic resonance spectroscopy (MRI/MRS) laboratory with an EPI sequence, and two participants were scanned in the Lucas Center of Stanford University with a spiral sequence. Because the participants were not required to perform any task during rest scans, the data acquired during such scans should be a good measurement of fMRI noise.

In contrast to the machine noise, the noise in the gray matter was typically maximum at the lowest measurable frequency (0.004 Hz) and declined progressively with the frequency, leveling out at higher frequencies. The decline with frequency is an important result because it implies that the noise-limiting BOLD signal becomes lower as the analysis frequency is increased, up to a certain value. The cortical noise amplitude is typically far larger than the extracephalic (hardware) noise at low temporal frequency. The difference can be as much as two log units, tending to converge at higher temporal frequencies. Note that this entire spectral range is well below 1 Hz, so we should not expect significant intrusion from heart pulses at ≥1 Hz. There might be some contamination from the respiration rate, however, because it should fall near 0.3 Hz. Nevertheless, we could not identify significant intrusion of cardiopneumatic noise, which would be realized as a spike in the power spectrum.

The noise characteristics specific to the cortex, or gray matter, may be compared with the results of the corresponding procedure after isolating the underlying white matter in the same regions of the brain during the same scans. The results (Fig. 4) are qualitatively similar to those for the cortex, seen in Fig. 3.

Because the noise spectrum is different inside and outside the head, it is evident that it consists of at least two components. We may analyze the component structure based on the simplifying assumption that there are two components that are additively independent: a component that derives from the dynamics of the BOLD activity (hemodynamic noise) and a component determined by thermal and other magnetic sources of noise in the scanner and its sensors (hardware noise). To obtain an uncontaminated estimate of the hemodynamic component of the noise, we make the assumption that the hardware noise component is uniform throughout the fMRI images (except in regions where ghost images from the head appear—any such regions were avoided for the extracephalic analysis). This assumption is justified by the flat spectrum for extracephalic noise and for the water phantom (Figs. 1, 2). Hence, the spectrum of the hardware noise should be $E(f) = h$, where h is a constant and f is the temporal frequency.

The cortical hemodynamic noise is assumed to have a uniform slope modeled by a power function $B(f) = bf^p$, where f is the frequency and p is the spectral slope. This power function corresponds to a straight line on a log–log plot and was termed a "1/f-like characteristic" by Bullmore et al. [5]. We assume that the hemodynamic noise and the hardware noise are independent from each other and combine by

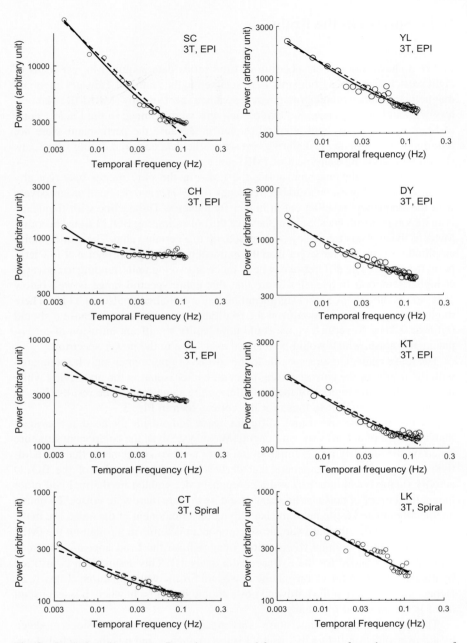

Fig. 3. Cortical resting spectra. Data show temporal frequency spectra from the gray matter of six observers. *Solid curve*, two-component model fits to the amplitude spectra. This model captures the systematic variation of the data over temporal frequency. *Dashed lines*, the 1/f-like function (f-p) fits

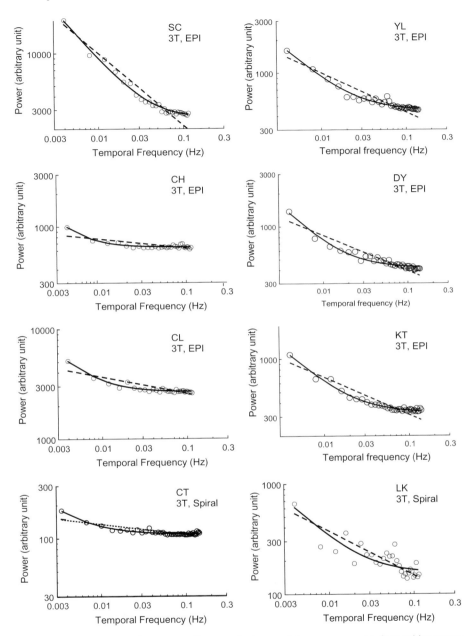

Fig. 4. White-matter resting spectra. Data are the temporal frequency spectra from white matter in the same scan. *Solid curve*, two-component model fits the amplitude spectra

adding their variances. Hence, in the cortex the noise spectrum, after the two sources combined, is

$$C(f) = [a \times E(f)^2 + b \times B(f)^2]^{0.5}$$

The fits for the two-component model are shown in Figs. 3 and 4 as the continuous curves for gray and white matter, respectively. The fit of this model is much better than 1/f-like noise alone (dashed lines in Figs. 3 and 4). On average, the sum-of-squares error (SSE) for the two-component model is three times better than that for the simple 1/f-like model for the gray matter and more than five times better for the white matter. Even considering the extra parameter, the reduction of SSE for the two-component model is highly significant: $F(8,414) = 46.72$ ($P < 0.0001$) for gray matter and $F(8,414) = 57.39$ ($P < 0.0001$) for white matter, pooled across participants. The slopes of the decreasing part of the spectrum varied from -0.45 to -0.92 across participants, with a mean of -0.65 ± 0.18 for gray matter and from -0.68 to -0.89 with a mean of -0.69 ± 0.25 for white matter. These slopes, which are not significantly different from one another, are about double those of the 1/f-like model fit, which has mean slopes of -0.36 ± 0.18 for gray matter and -0.30 ± 0.19 for white matter.

Mechanisms of Spectral Slopes

For the present data, the spectrum is almost flat at high temporal frequencies for most observers. This flat spectrum implies a white noise characteristic in this frequency region. In contrast to the random walk, which is a "long memory" process, white noise is memoryless. That is, the history of the noise-generating mechanism does not affect current responses. Our analysis supports the idea that the measured noise derives from a combination of a long-memory component and a memoryless component of the overall noise.

The major issue now becomes the interpretation of the hemodynamic component. The decrease in noise power with temporal frequency is reminiscent of what is commonly called "1/f noise" for the amplitude spectrum (which is the square of the amplitude spectrum), or "pink" noise. In fact, the average log–log slope for our data was -0.67, somewhat shallower than the full 1/f property despite removal of the white noise component. For 1/f behavior of the amplitude spectrum, there are two classic explanations: random-walk behavior or white noise followed by a first-order filter. Random-walk behavior implies that the generating system randomly increases or decreases from its current position by a fixed amount during each (arbitrary) time interval. A random walk mechanism would have an unlimited 1/f spectrum down to whatever was the lowest measured frequency. The random walk is characteristic of increments governed by coin-flipping and of systems such as stock markets; and it is independent of the amplitude distribution governing the step size.

The other type of mechanism, the first-order filter, has asymptotic 1/f behavior for frequencies much higher than its corner frequency but an asymptotic slope of zero at low temporal frequencies. The only way to fit the present temporal spectra with such a filter would be to assume that the f^{-p} slope was governed by a process whose "corner" fell in the range of the measured frequencies in Fig. 3, providing an empirical slope that was shallower than 1/f. However, the data show no tendency to level off at the lowest frequency (0.004 Hz), so we assume that the relevant model for the physiological noise component is, in fact, the f^{-p} model.

Further analysis shows that the f^{-p} model of cortical BOLD noise is incompatible with the known blood dynamics in two ways. The blood dynamics relevant to the fMRI assay are characterized by the BOLD signal properties, which have been best analyzed by Buxton et al. [12] and Glover [10]. A representative fit of their models is a biphasic function with shallow resonance and a time constant of about 10 s. If this model characterizes the hemodynamic impulse response, it must necessarily be the filter that defines the hemodynamic limit on the noise spectrum (assuming that the noise generator has a flat spectrum). The frequency spectrum of such a hemodynamic filter model, as shown later, has a band-pass property that is obviously different from the noise spectrum. Zarahn et al. [4] made a similar point with respect to the 1/f fit to their data.

We therefore conclude that the most likely candidate to explain the hemodynamic noise spectrum is a separate f^{+p} property of the noise generation process rather than it being a reflection of the hemodynamic response properties. This separate noise generation process may operate through an extended series of filters or be a kind of constrained random walk process. Because the distribution on each voxel is accurately Gaussian [2], the falloff with frequency cannot be due to a nonlinearity in the amplitudes of the signal modulations but must be inherent in the time domain. At frequencies above about 1 Hz, this analysis suggests that the BOLD noise would conform to the f^{-p} property, flattening out when it reaches the level of the extracephalic noise. Our data suggest that the extracephalic-to-cortical noise ratio varies idiosyncratically among observers (Figs. 3, 4), so the frequency at which flattening occurs should also be idiosyncratic.

It is of interest that other aspects of brain processing also show a steep decline of noise amplitude with temporal frequency. Tyler et al. [13], for example, reported that the noise spectrum of the electrical brain activity fell with a log–log slope close to −1 over the range of 1–100 Hz. To the extent that both the scalp electrical potential and the BOLD hemodynamic response derive from slow electrochemical potentials in the neural cell bodies and their processes, it is not surprising to find a continuum of behavior in these activities. The implication of this joint result is that the neural potentials exhibit an approximation to 1/f behavior over more than a four-decade range of temporal frequencies, from 0.004 to 100 Hz.

Signal

Signal Spectrum

The signal spectrum of BOLD activation to certain sensorimotor or cognitive tasks may be obtained with a block design experiment. A typical block design fMRI

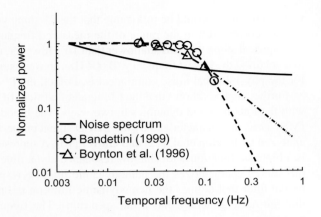

Fig. 5. Amplitude spectra for task-induced response in the motor (*open circles* and *dashed curve*) and the visual cortex (*open triangles* and *dash-dot curve*) and for the noise in the gray matter (*solid curve*)

experiment has the participants alternately perform two tasks several times. The two tasks, ideally, differ in only one neuronal function. The experimenters then contrast the BOLD activation difference to these two tasks. If one systematically changes the temporal frequency of the alternation and observes how BOLD activation contrast changes with the temporal frequency, one can obtain the BOLD contrast sensitivity function (i.e., the BOLD contrast versus temporal frequency function) for the neuronal function in question.

Bandettini [7] measured the BOLD contrast sensitivity function for finger movement in the motor cortex. The BOLD contrast increases with the decrease of the temporal frequency. As shown in Fig. 5, we were able to fit a low-pass filter function (dashed curve) to his data (open circles). The data were derived from Bandettini's Figure 19.3 [7]. The noise spectrum (solid curve) is plotted in the figure as a comparison. The best-fitting filter function was second order (i.e., a two-pole filter). Boynton et al. [8] reported the BOLD contrast sensitivity function for viewing an 8-Hz counter-phase flickering checkerboard in the primary visual cortex. The open triangles in Fig. 5 were taken from the GMB data in Figure 6 of Boynton et al. [8]. Again, the BOLD contrast sensitivity function (dash-dot curve in Fig. 5) is well fit with a low-pass filter function. For these data, the best-fitting function was first order (i.e., a one-pole filter).

The hemodynamic response is essentially linear under most circumstances [8]. This implies that (1) the power spectrum of BOLD activation is the Fourier transform of the impulse response function [called hemodynamic response function (HRF) in this context] of the system; and (2) the fMRI time series is a convolution between HRF and the experimental sequence designated by the experimenter. These properties allow us to estimate the signal power spectrum from a vast body of time-domain data in the literature.

The shape of the hemodynamic response function is well known. Although there is disagreement about whether there is an initial dip 1–2 s [14] after the stimulus onset, it is generally agreed that the BOLD activation reaches its peak at around

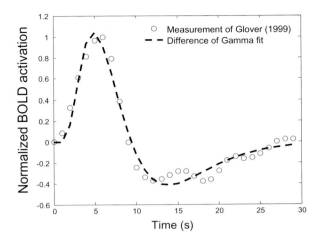

Fig. 6. Difference of gamma fit to a hemodynamic response measured by Glover [10]

4–6 s and is followed by a shallow undershoot that lasts up to 15–20 s after the stimulus onset. The theoretical hemodynamic impulse response functions show a similar biphasic form [12, 15].

Figure 6 shows an example of the BOLD response after the onset of a short (1 s) auditory stimulus measured in the temporal cortex by Glover [10]. This function, we found, can be well fit by a difference of two gamma functions defined by

$$dg(x) = w_1 (x/\alpha_1)^{\beta_1} e^{-x/\alpha_1} - w_2 (x/\alpha_2)^{\beta_2} e_2^{-x/\alpha_2}$$

where w_i, α_i, and β_i, $i = 1, 2$ are free parameters. Among those parameters, a_I determines the position of the peak of the i-th component in time, b_I determines the steepness of response change in the i-th component, and w_i is a parameter of an arbitrary unit that determines the maximum response change of the i-th component.

As is shown by the smooth dashed curve in Fig. 6, this function incorporates sufficient parametrization to capture all the features of the responses [cf. 16]. We then Fourier-transformed this fitted response function to derive the predicted power spectrum of the BOLD activation as the dashed curve in Fig. 7. Again, the typical noise spectrum (solid curve) is plotted in the figure as a comparison. In contrast to the power spectrum obtained in the frequency domain, the signal power spectrum in the time domain shows a band-pass property with maximum power at 0.04 Hz.

The HRF can be derived by deconvolving the fMRI time series data and the stimulus sequence designated by the experimenter. Chen et al. [16] did this analysis for their visual experiment and fit the derived HRFs in the visual cortex with the difference in gamma functions. Here, we took the median of the difference in gamma parameters of the six participants in the study of Chen et al. [16] and computed a representative HRF of the visual cortex. We then Fourier-transformed this

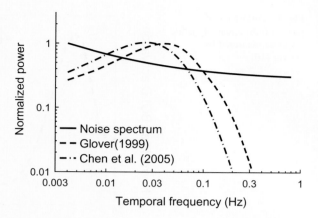

Fig. 7. Amplitude spectra for task-induced response in the temporal cortex (*dashed curve*) and the visual cortex (*dash-dot curve*) and for the noise in the gray matter (*solid curve*)

fitted HRF to obtain the power spectrum of the BOLD activation as the dash-dot curve in Fig. 7. This curve also shows a band-pass property, with maximum power occurring at 0.032 Hz. This peak frequency corresponds to a 31-s period of task switching in a block design, or a 15.5-s epoch length. The power spectrum estimated from Glover [10] showed a peak frequency at 0.04 Hz, corresponding to a 12.5-s epoch length in a block design. The difference between the two studies, 3 s, is even smaller than the TR (repetition time) of the 3.5 s used by Chen et al. [16] and would not be detectable in most block design experiments.

There is also a discrepancy between the power spectra derived from the time domain studies (which show a band-pass property as in Fig. 7) and those directly measured in the frequency domain (which show a low-pass property as in Fig. 5). The band-pass property of the power spectrum is expected based on the fact that the HRF shows a biphasic shape, which implies attenuation of the system response to a longer stimulus and tuning to an optimal temporal frequency. The question is why the power spectrum measured in the frequency domain did not show attenuation at the low frequencies, implying a monophasic HRF. The reason of this discrepancy remains unresolved. One possible reason is that the undershoot is attributable to a nonlinearity in the system, such as recovery from early adaptation for long stimulation that is equated across frequency in the frequency measurement paradigm.

Signal-to-Noise Ratio

Although there are discrepancies in the signal power spectra from various studies, the signal-to-noise (SNR) shares largely similar properties. Figure 8 shows the typical SNR as a function of temporal frequency assuming the typical noise parameters derived from Fig. 3. The solid curve is the SNR computed by dividing the signal spectrum from the temporal cortex to an auditory stimulation [10]; the

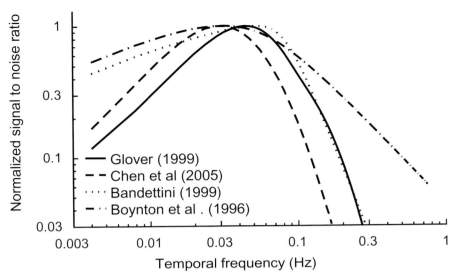

Fig. 8. Ratio of the task-induced response (signal) spectrum from four studies (see keys) and the noise spectrum

dashed curve, from the visual cortex to a visual stimulation [16]; the dotted curve, from the motor cortex during finger movement [7]; and the dash-dot curve, from the visual cortex to a visual stimulation [8]. Despite all these differences in brain areas and tasks, all the SNR curves show a band-pass property. The peak frequency ranges from 0.028 and 0.032 Hz in the two visual cortex studies [8, 16] to 0.04 Hz in the temporal cortex study [10] to 0.052 Hz in the motor cortex study [7]. This implies that, with a block design, a period of around 20–36 s or paired test and null epochs of 10–18 s should be the optimal stimulation rate.

Acknowledgments This work was supported by NIH/NEI grants 7890 and 13025 to C.W.T. and NSC grant 93-2752-002-H-007-PAE to C.C.C. Thanks go to Heidi Baseler for data collection of the spiral scans at Stanford University and to Chen Der-Yo and Liu Chia-Li for data collection of the spiral scans at National Taiwan University.

References

1. Ogawa S, Lee TM, Ray AR, et al (1990) Brain magnetic resonance imaging with contrast dependent on blood oxygenation. Proc Natl Acad Sci USA 87:9868–9872.
2. Chen CC, Tyler CW, Baseler HA (2003) The statistical properties of BOLD magnetic resonance activity in the human brain. Neuroimage 20:1096–1109.
3. Aguirre GK, Zarahn E, D'Esposito M (1997) Empirical analyses of BOLD fMRI statistics. II. Spatially smoothed data collected under null-hypothesis and experimental conditions. Neuroimage 5:199–212.

4. Zarahn E, Aguirre GK, D'Esposito M (1997) Empirical analyses of BOLD fMRI statistics. I. Spatially unsmoothed data collected under null-hypothesis conditions. Neuroimage 5:179–197.
5. Bullmore E, Long C, Suckling J, et al (2001) Colored noise and computational inference in neurophysiological (fMRI) time series analysis: resampling methods in time and wavelet domains. Hum Brain Mapp 12:61–78.
6. Pfeuffer J, Van de Moortele PF, Ugurbil K, et al (2002) Correction of physiologically induced global off-resonance effects in dynamic echo-planar and spiral functional imaging. Magn Reson Med 47:344–353.
7. Bandettini PA (1999) The temporal resolution of functional MRI. In: Moonen CTW, Bandettini PA (eds) Functional MRI. Springer, Heidelberg, pp 205–220.
8. Boynton GM, Engel SA, Glover GH, Heeger DJ (1996) Linear systems analysis of functional magnetic resonance imaging in human V1. J Neurosci 76:4207–4221.
9. Kwong KK, Belliveau JW, Chesler DA, et al (1992) Dynamic magnetic resonance imaging of human brain activity during primary sensory stimulation. Proc Natl Acad Sci USA 89: 5675–5679.
10. Glover GH (1999) Deconvolution of impulse response in event-related BOLD fMRI. Neuroimage 9:416–429.
11. Stehling MK, Turner R, Mansfield P (1991). Echo-planar imaging: magnetic resonance imaging in a fraction of a second. Science 254:43–50.
12. Buxton RB, Wong EC, Frank LR (1998) Dynamics of blood flow and oxygenation changes during brain activation: the balloon model. Magn Reson Med 39:855–864.
13. Tyler CW, Apkarian P, Nakayama K (1978) Multiple spatial-frequency tuning of electrical responses from human visual cortex. Exp Brain Res 33:535–550.
14. Hu X, Le TH, Ugurbil K (1997) Evaluation of the early response in fMRI in individual subjects using short stimulus duration. Magn Reson Med 37:877–884.
15. Aubert A, Costalat R(2002) A model of the coupling between brain electrical activity, metabolism, and hemodynamics: application to the interpretation of functional neuroimaging, Neuroimage 17:1162–1181.
16. Chen CC, Tyler CW, Liu CL, et al (2005) Lateral modulation of BOLD activation in unstimulated regions of the human visual cortex. Neuroimage 24:802–809.

Magnetoencephalography: Basic Theory and Estimation Techniques of Working Brain Activity

Yumie Ono[1] and Atsushi Ishiyama[2]

Summary

It is widely accepted that magnetoencephalography (MEG) is a promising tool for investigating human brain activity in good temporal and spatial resolution. However, the use of MEG is currently quite limited, mostly due to excessive diversity in localization techniques to estimate regional brain activity using MEG data. Because source localization in MEG is an ill-posed problem, an adequate localization technique varies depending on the task design and the timing of interest in MEG signals, which sometimes confuses investigators when choosing an analysis technique or comparing the results with those obtained with other modalities, such as functional magnetic resonance imaging. This chapter reviews the introductory theories and applications of currently available MEG source localization techniques as well as principles of MEG signals and its measurement for beginners and possible future MEG users. The physiological and mathematical backgrounds of cerebral MEG source are briefly introduced followed by the technical requirements for MEG data acquisition. Modern localization techniques for inverse problem solving with MEG, from a simple dipole model to an underdetermined type of norm estimation or spatial filter technique, are thoroughly described.

Key words Magnetoencephalography, Inverse problem, Source localization, Dipole model, Spatial filter

Introduction

Modern brain function imaging tools, such as functional magnetic resonance imaging (fMRI), positron-emission tomography (PET), electroencephalography (EEG), and magnetoencephalography (MEG), have been dramatically improved

---

[1]Department of Physiology and Neuroscience, Kanagawa Dental College, 82 Inaoka-cho, Yokosuka, Kanagawa 238-8580, Japan

[2]Department of Electrical Engineering and Bioscience, School of Science and Engineering, Waseda University, 3-4-1 Okubo, Shinjuku-ku, Tokyo 169-8555, Japan

over decades. They are now widely employed in clinical, neuroscience, medical engineering, and psychological studies using human subjects. When the utilization of these brain imaging tools is researched regarding the number of publications in PubMed, the key words fMRI, PET, EEG, and MEG are found in 63 096, 9512, 55 638, and 1597 articles, respectively, in combination with "brain" (the search results were obtained in February 2007).

The volume of publications related to MEG is only 2.5% of that with fMRI. Why is the utilization of MEG still quite limited? One possible reason is the lack of versatile visualization software of regional brain activity for MEG data. In the case of fMRI, statistical parametric mapping (SPM) [1, 2] is a gold standard among researchers to specify regional brain activities at arbitrary voxels in the brain. However, for MEG, diverse estimation methodologies have been proposed by various researchers, but as yet there has been no unified estimation software. This is because the estimation of brain regional activity from MEG is an "ill-posed" inverse problem; that is, we cannot specify the brain activation pattern uniquely and so have to choose an optimal method based on physiological and mathematical assumptions. However, when an estimation method is adequately chosen, MEG will become a powerful tool for studying human brain function with good temporal (milliseconds) and spatial (several millimeters) resolution; and it will contribute to further investigation of brain function in combination with other modalities. Here, the theories and applications of currently available MEG source localization techniques are introduced for beginner and possible future MEG users with descriptions of the physiological, mathematical, and technical backgrounds of MEG measurements.

Generation of MEG Signals

A sensory stimulus activates the corresponding sensory area of the cerebral cortex. The activation is given as a small current source through the neurons, which is called a primary current. At the same time, the volume currents are induced in the surrounding medium to cancel the primary current, as shown in Fig. 1a. The current density of the primary current is considerably higher than that of the volume current because the primary current is restricted in a narrow intracellular area whereas the volume current is widely dispersed in an extracellular area of the brain. Therefore, the MEG signal is mainly caused by the primary current, even though the magnetic field is generated from both the primary and volume currents. The magnetic field generated from the primary current by Biot-Savart's law is indicated in Fig. 1a.

The excitatory postsynaptic potential (EPSP) in pyramidal neurons of the cerebral cortex is thought to be a source of the primary current [3, 4]. Owing to the long EPSP duration of more than 10 ms, the EPSP-evoked magnetic fields from synchronous excitation of neighboring pyramidal neurons can be readily superimposed on each other to generate a MEG signal that is detectable over the scalp

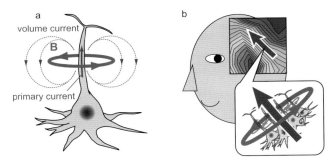

Fig. 1. Magnetic field generated by primary current. **a** Excitatory postsynaptic potential (EPSP)-related intracellular ion current (primary current) and the magnetic field induced. **b** Magnetoencephalography (MEG) is observed when the individual magnetic fields are superimposed by simultaneous excitation of a number of cerebral pyramidal cells

(Fig. 1b). Practically, the biomagnetic signal is generated from activation of ~1 cm^2 of the cortex area, and the estimated amplitude of the neuronal current is less than 0.4 nAm/mm^2 [3].

Because the length of the neuronal primary current is negligible compared to the volume of the head, the source of the MEG signals can be approximated as a current dipole in the head assumed to be a magnetically homogeneous spherical conductor. It is a widely used assumption for source localization of MEG signals because of the computational simplicity and the reasonable location accuracy of the estimated current source. With this model, only the tangential component of a current dipole to the surface of the brain is sensitive to the magnetic field outside the conductor (for details, see [5, 6]). It suggests that MEG signals observed over the scalp mainly represent the neuronal activity in the sulcus and that the activity in the gyrus is less contributive to the MEG signals. However, MEG is still effective because the target areas of the MEG investigation are mostly located in the sulcus area, such as the primary sensory areas.

Detection of MEG Signals

MEG System

The setup of a MEG system and an example of auditory evoked MEG are shown in Fig. 2. The system is typically composed of the following instruments.

Magnetic Sensors. The amplitude of magnetic signals observed over human scalp is in the range of several tens to hundreds femto Tesla (fT), which is equivalent to one hundred millionth of the geomagnetic field intensity. The superconducting quantum interference device (SQUID) [7] is the only sensor that can detect such a small

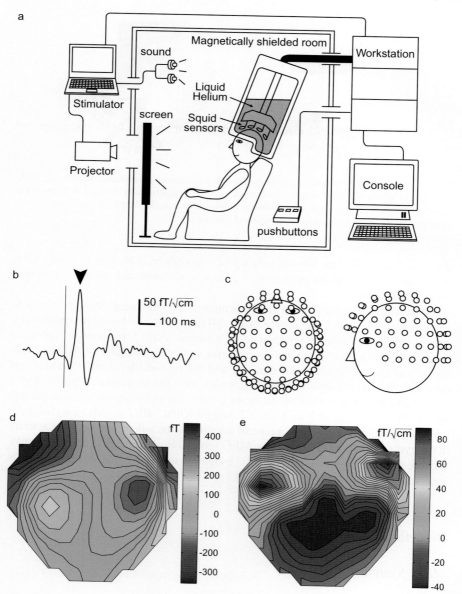

Fig. 2. Systems and responses of MEG. **a** A MEG system. Typically, the system consists of a dewar with magnetic sensors, a magnetically shielded room, workstations, and sensory stimulators. **b** Example of an auditory evoked magnetic field measured over the auditory area. The *vertical line* shows the onset of a 1-kHz tone burst sound, and the *arrowhead* indicates the time point of maximum auditory response. **c** Arrangement of 102 channels of magnetic sensors (Elekta Neuromag Oy, Vectorview system) over the scalp. **d, e** Distribution of the magnetic field of the auditory response at the time point indicated in **b**, which is represented by the values of the magnetometer directly (**d**) and by the root-mean-square values of orthogonal planar gradiometers (**e**), respectively

magnetic field. Several tens to hundreds of channels of SQUID sensors, working in ultra-low temperature, are arranged in a helmet-shaped dewar filled with liquid helium.

Workstation. Adjustment and control of SQUID sensors, data acquisition, stimulators, and signal processing are available.

Electromagnetically Shielded Room. The altenating current (AC) component of external magnetic noise is prevented by a shield with high electrical conductivity. A multilayer shield of high-magnetic permeability metals is equipped to prevent SQUID sensors from picking up environmental magnetic noise as well. The typical shielding performance is around 60 dB at 1 Hz. Some of them are further equipped with an active shielding system [8], which produces uniform and/or spatial gradient magnetic fields to cancel ambient noise. Instruments for sensory stimulation arranged inside a shield room (e.g., video screen, auditory stimulators, push buttons) should be made of nonmagnetic material. For more detailed description of the instrumentation for MEG, see Hämäläinen et al. [9, section V].

Magnetic Noise in MEG and Its Reduction

The detection of regional brain activity (i.e., the accuracy of the MEG inverse problem) is directly related to the signal-to-noise ratio of the measured data. Because the environmental magnetic field is much stronger than the cerebral magnetic field, as shown in Fig. 3, deletion of external magnetic noise is essential. The typical noise sources for MEG data are classified into the following areas.

External Noise

The movement of magnetic materials outside the laboratory (e.g., the operation of vehicles and elevators) generates low-frequency magnetic field fluctuation. The surrounding electronic devices, including the MEG workstations, are possible sources of electromagnetic noise at the AC power line frequency. Even muscular electrical activities of a subject (e.g., breathing, heartbeats) generate a magnetic field that can easily get mixed in with MEG signals. The external noise is a collective term describing these magnetic fields that come from outside of the head. MEG data are often collected in a magnetically shielded room using a notch filter to prevent the external noise. Additionally, most MEG systems adopt a gradiometer as a pickup coil of magnetic field to prevent external noise. Pickup coils are generally classified into two types: the single-loop type of magnetometer, which directly detects the average of the magnetic fields linking a single coil (Fig. 4a), and the twisted-loop type of gradiometer, which detects the spatial gradients of adjacent magnetic fields (Fig. 4b,c). Because the magnetic field produced by a current dipole decreases in inverse proportional to the cube of the distance, the

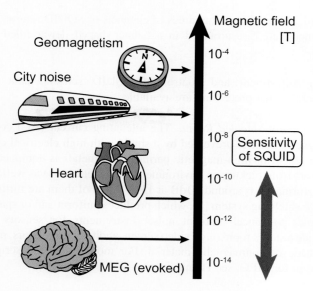

Fig. 3. Amplitudes of magnetic fields due to biomagnetic and noise sources. *SQUID*, superconducting quantum interference device

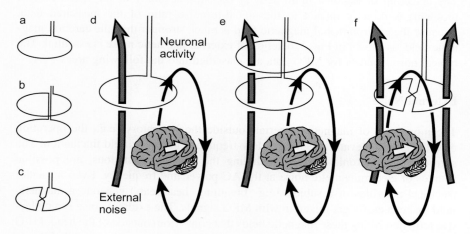

Fig. 4. Variety of pickup coils and principles of noise cancellation. **a** Magnetometer. **b** Symmetrical series axial gradiometer. **c** Series planar gradiometer. **d–f** Fields of uniform external noise (*gray arrows*) and inhomogeneous neuronal activity (*black arrows*) are both sensed by the pickup coil of the magnetometer (**d**). However, the external noise is canceled when gradiometers (**e, f**) are used

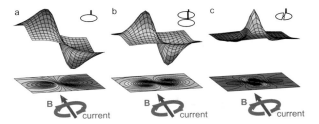

Fig. 5. Differences in the observed field patterns depending on the design of the pickup coils. Upper surface plots and the middle contour plots indicate the field pattern observed by a magnetometer (**a**), axial gradiometer (**b**), and planar gradiometer (**c**). The magnetic fields were calculated with Sarvas' equation [12] with a semi-infinite plane conductor model in which a tangential current dipole is assumed

external magnetic field from a distant dipole-like source is relatively uniform at pickup coils. It is therefore possible to cancel out the external noise with specially designed gradiometers in most cases.

As shown in Fig. 4b, an axial gradiometer consists of a lower loop of pickup coil and an upper loop of compensation coil, which have identical areas but are placed the opposite way around to each other. The external magnetic field would be canceled by the compensation coil because its spatial distribution is considered to be nearly uniform at both coils. In contrast, the magnetic field from a nearby cerebral source distributes inhomogeneously around these coils, so it is selectively detected by the gradiometer. Another type is a planar gradiometer (Fig. 4c), whose pickup coil is twisted at a center of the loop, detecting the horizontal gradient of magnetic fields. Among the currently available commercial systems, axial gradiometers are adopted in the CTF MEG systems (VSM MedTech, Coquitlam, BC, Canada), and the combinations of a magnetometer and two orthogonal planar gradiometers at each measurement site are adopted in the Elekta MEG systems (Elekta Neuromag Oy, Helsinki, Finland). Note that the magnetic field pattern would be observed differently depending on the design of the pickup coils (Figs. 2d,e, 5). Magnetic influx and efflux peaks are observed across the cerebral source position when the magnetic field is detected by magnetometers or axial gradiometers. Meanwhile, a maximum response is given just above the cerebral source by planar gradiometers.

System Noise

The signal-to-noise ratio of stimulation-evoked MEG is usually <1 even though the data were obtained by gradiometer in a magnetically shielded room. Other than the external noise, the MEG signals still contain system noise, thermal noise from SQUID itself, and brain noise. Because the system noise is mostly Gaussian white noise, averaging is widely used effectively to attenuate the noise level (Fig. 6). Under the assumption that the signal component is reproducible and independent of a random noise component, n times of averaging improves the signal-to-noise

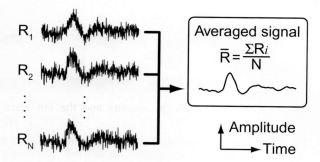

Fig. 6. Signal averaging for noise cancellation. A series of time-locked raw data are collected and averaged over epochs. It is assumed that the signal component is identical over epochs, and the noise component is statistically noncorrelated to the stimulus

ratio by $\sqrt{n}$ times. Several tens to thousands times of signal averaging is applied in the practical MEG measurements. Averaging is the most generally accepted noise reduction technique for any kind of stimulation-evoked MEG measurements. However, care should be taken during the experimental task to obtain reproducible responses when averaging is applied.

Brain Noise

Brain noise refers to the magnetic fields from the head area in which an experimenter is not interested (e.g., spontaneous MEG signals during the evoked MEG measurement) and the magnetic field generated by facial muscle movement (e.g., eye movements or blinks). Deletion of spontaneous MEG signals by digital filters is difficult because they show up irregularly and vanish depending on the physical and psychological condition of the subject and their frequency band is similar to that of evoked signals. However, signal averaging somewhat attenuates spontaneous signals that are independent of the stimulation task. The blink-related distortion of magnetic fields is also severe because the eye muscle is very close to the magnetic sensors. The blinks during measurement can be detected by electrooculography. Alternatively, the response contaminated by a blink artifact can be manually or automatically deleted from MEG signals because it is thousands of times the strength of a typical evoked MEG.

Integration to Morphological MRI

The estimated current source is generally represented as the position and moment of a current dipole at a certain time window, which does not provide an anatomical location in the subject's individual brain. Superimposition of the estimated current dipoles on the subject's morphological MRI images is necessary to confirm the physiological location of the activated area in the brain.

Markers are attached to several cephalic landmarks when MRI scans are obtained for the purpose of MEG–MRI integration. We usually attach vitamin E liquid capsules to the nasion and binaural tragus of a subject. When the subject participates in a MEG experiment, three or four head position indicator (HPI) coils are attached to the subject's face. The positional relation of the same landmark points in MRI and HPI coils are digitized and stored in the workstation in advance. Just before data acquisition, a weak current is applied to each HPI coil, and the positions of the HPI coils are detected by the generated magnetic field patterns. The relative positions between magnetic sensors and the landmark points are consequently calculated using the positional relation between the HPI coils and the landmark points. Because we cannot detect the head position during data acquisition, subjects are encouraged to keep their head position as stable as possible once the HPI coil positions are detected. Shifting the head position during data acquisition results in poor accuracy in source localization.

Source Localization of MEG

The inverse problem in MEG is used to estimate the cerebral current sources from the distributions of magnetic field over the scalp, which is often called "source localization." When the neuronal activities are assumed to localize spatially and distribute sparsely in the brain, an overdetermined type of inverse problem is adopted. It assumes that a single current dipole or several current dipoles exist in a head conductor (Fig. 7a). This assumption is well suited to a primary sensory response, such as a short-latency somatosensory and primary auditory evoked field. However, it is not suitable for brain activity reflecting spontaneous or higher cognitive function [10, 11] in which simultaneous activation of multiple or widely distributed cortical areas is considered (Fig. 7b). An underdetermined type of inverse problem is considered an effective tool for analyzing such a complicated neuromagnetic field. In this case, the brain is divided into a number of mesh points, and the current density at each point is calculated by means of the norm estimation or the spatial filter.

Dipole Approximation of Cerebral Source

As described in the section of MEG generation, a current dipole **Q** at r_Q is a good approximation of point-like primary current $\mathbf{J}^p$:

$$\mathbf{J}^p(\mathbf{r}) = \mathbf{Q}\delta(\mathbf{r} - \mathbf{r}_Q) \tag{1}$$

where $\delta(\mathbf{r})$ is the Dirac delta function. The current dipole is thought to represent an aggregation of the primary currents from synchronous activation of pyramidal cells in the cortex. If the dipole approximation is valid (i.e., the distribution of measured magnetic field shows a dipolar pattern), we can precisely determine the position, direction, and moment of cerebral current source **Q**. The determined

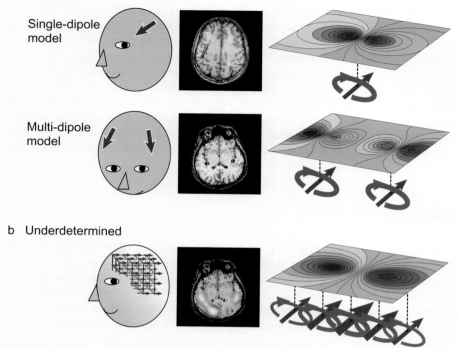

Fig. 7. Variety of inverse problem solutions in source localization. **a** An overdetermined inverse problem is adopted when the observed magnetic field is spatially localized, as shown in the contour maps (*right*). The estimation is given as positions and moments of current dipoles, which are usually superimposed on a subject's magnetic resonance (MR) image for the anatomical evaluation of signal source (*center*). **b** An underdetermined inverse problem is adopted when the widely distributed signal sources are assumed. The estimation is given as the amplitude of current sources at each mesh points defined in the brain. Anatomical evaluation with MR images is also inevitable to acquire the precise active location in the brain

current source is generally called an equivalent current dipole (ECD). The magnetic field **B** at arbitrary position **r** outside the conductor in a spherical head model was given by Sarvas [12] as follows:

$$\mathbf{B}(\mathbf{r}) = \frac{\mu_0}{4\pi} \frac{F\mathbf{Q} \times \mathbf{r}_Q - (\mathbf{Q} \times \mathbf{r}_Q \cdot \mathbf{r})\nabla F(\mathbf{r},\mathbf{r}_Q)}{F(\mathbf{r},\mathbf{r}_Q)^2} \qquad (2)$$

where

$$\mathbf{a} = (\mathbf{r} - \mathbf{r}_Q) \; a = |\mathbf{a}|, \quad r = |\mathbf{r}|$$

$$F(\mathbf{r},\mathbf{r}_Q) = a(ra + r^2 - \mathbf{r}_Q \cdot \mathbf{r})$$

and

$$\nabla F(\mathbf{r},\mathbf{r}_Q) = (r^{-1}a^2 + a^{-1}\mathbf{a}\cdot\mathbf{r} + 2a + 2r)\mathbf{r} - (a + 2r + a^{-1}\mathbf{a}\cdot\mathbf{r})\mathbf{r}_Q$$

Equation 2 indicates that the radial current source is silent outside the conductor, which suggests that the magnetic field detected over the scalp comes mainly from tangential primary current (i.e., brain activity at the sulci).

Overdetermined Inverse Problem

When the measured magnetic field is approximated by a small number of current sources (less than the number of magnetic sensors), the source localization technique is called an overdetermined inverse problem. MEG source localization using current dipole approximation has now been widely adopted, as responses from the primary sensory area are highly localized and well explained with this model. For example, the somatosensory homunculus from face to toes [13, 14] or the visual retinotopy [15, 16] can be determined noninvasively with a spatial resolution of 3–5 mm in the individual brain. Clinically, the ECD source localization is widely utilized to determine sensory areas before surgically removing a tumor or epileptic cortex [17, 18]. A single-dipole model (e.g., somatosensory evoked responses) or a two-dipole model (e.g., auditory evoked responses) is usually adopted to localize ECDs related to the primary sensory responses. Another major clinical application is the determination of epileptic foci from spontaneous MEG of a patient [19, 20]. A large amount of basic neuroscience research has been done as well to pursue somatosensory [21], pain [22], motor [23, 24], visual [16], auditory [25], gustatory [26], and olfactory [27] sensory systems of a human brain.

Single-Dipole Model

Geometric Estimation. A tangential current dipole in a semi-infinite plane conductor generates two peaks of magnetic influx and efflux at the measuring surface, as shown in Fig. 5. If the distance between the peaks is described as d, the current dipole location is at the middle of these two peaks horizontally with the depth of $d/\sqrt{2}$. We can simply guess the source position from the magnetic field distribution, even though it is not as precise in the spherical measurement plane of the MEG helmet.

Explorative Estimation. When the position and moment of a current dipole in a spherical model are given, the magnetic field at each magnetic sensor can be calculated from the forward problem in Eq. (2). The aim of the source localization is to determine the position and moment of a current dipole, which best explain the measured magnetic field. The mean squared error between the measured magnetic field Bm^i ($I = 1, \ldots, N$; where N is a number of magnetic sensors) and the calculated magnetic field Bc^i is described as follows.

$$e = \sqrt{\frac{\sum (Bm^i - Bc^i)^2}{\sum (Bm^i)^2}} \qquad (3)$$

The position and moment of a dipole are iteratively sought to minimize the evaluation function e. The steepest descent method, the simplex method, and the stochastic algorithm—such as a genetic algorithm (GA) or simulated annealing (SA)—are the representative algorithms of the explorative estimation. When the steepest descent method or the simplex method is used, estimation of the same data should be repeated several times with different initial values to prevent a local minimum solution derived from the noise in practical MEG data. The GA and SA have less chance of getting into a local minimum solution irrespective of the initial values. GA stands on the idea of evolutionary biology (e.g., inheritance, mutation, selection, crossover) in each iterative step. SA adopts an additional parameter of temperature gradient, which reflects the iteration count, with which the range of search space is narrowed and the probability of accepting variables that makes the evaluation function worse are determined. However, the accuracy of GA or SA depends on the number of iterative calculations; and a relatively long computation time is needed.

Multiple Dipole Model

The multiple-dipole model is an expansion of the single-dipole model. The number of current dipoles is defined in advance, and the explorative estimation methods introduced for the single dipole model are adopted to find appropriate positions and moments of the current dipoles that minimize the evaluation function. The number of current dipoles to estimate is usually determined from the spatial distribution of the magnetic field and/or from an a priori physiological assumption on the task. Several techniques have been introduced to estimate multiple neuronal sources without foreknowledge of the number of current dipoles, such as principal component analysis-based multiple signal classification (MUSIC) [28] and fMRI-aided MEG source localization in which fMRI-active regions are used to confine the search space of current dipoles during source localization [29, 30].

Underdetermined Inverse Problem

The underdetermined inverse problem estimates the moments of the current dipoles at each mesh point fixed in a head conductor. The contribution of a dipole $\mathbf{Q}^j = q_x^j \mathbf{e}_x + q_y^j \mathbf{e}_y + q_z^j \mathbf{e}_z$ ($\mathbf{e}_\zeta : \zeta = x,y,z$ are the unit vectors) at position $\mathbf{r}$ ($j = 1, \ldots,$

M, where M is a number of mesh points) to the magnetic field $\mathbf{B}(\mathbf{r}^i,\mathbf{r}_Q^j)$ measured at $\mathbf{r}^i$ ($I = 1, \ldots, N$, where N is a number of magnetic sensors, $N < M$) is derived from Eq. (2) as the following linear relation

$$\mathbf{B}(\mathbf{r}^i,\mathbf{r}_Q^j) = l_x^{i,j} \cdot q_x^j \mathbf{e}_x + l_y^{i,j} \cdot q_y^j \mathbf{e}_y + l_z^{i,j} \cdot q_z^j \mathbf{e}_z = \mathbf{L}^{i,j} \cdot \mathbf{Q}^j \quad (4)$$

where $l_\zeta^{i,j}$ are the constants determined by $\mathbf{r}_Q^j$ and $\mathbf{r}^i$. The matrix description of magnetic fields from ζ components of all the current dipoles is as follows.

$$\begin{bmatrix} B_\zeta(\mathbf{r}^1) \\ \vdots \\ B_\zeta(\mathbf{r}^i) \\ \vdots \\ B_\zeta(\mathbf{r}^N) \end{bmatrix} = \begin{bmatrix} l_\zeta^{1,1} & \cdots & \cdots & l_\zeta^{i,1} & \cdots & \cdots & l_\zeta^{1,M} \\ \vdots & \ddots & & \vdots & & & \vdots \\ \vdots & & \ddots & l_\zeta^{i,j} & & & \vdots \\ \vdots & & & \vdots & \ddots & & \vdots \\ l_\zeta^{N,1} & \cdots & \cdots & \cdots & \cdots & \cdots & l_\zeta^{N,M} \end{bmatrix} \cdot \begin{bmatrix} q_\zeta^1 \\ \vdots \\ \vdots \\ q_\zeta^j \\ \vdots \\ \vdots \\ q_\zeta^M \end{bmatrix} \quad (5)$$

$$\mathbf{B} = \mathbf{L} \cdot \mathbf{Q}$$

The **L** is called the lead field because it indicates the unit contribution of each current to the measured magnetic field. The solution in the undetermined inverse problem is to determine **Q**, which fills Eq. (5) with given **B** and **L**. The spatial resolution of the estimation depends on the density of mesh points. Practically, the head conductor is divided into hundreds to thousands mesh points with intervals of several millimeters. To decrease the computational time, the mesh points can be arranged selectively at the cerebral cortex with the aid of morphological MRI.

Norm Estimation

The simplest solution of Eq. (5) is the minimum-norm estimation (MNE), in which the root mean square of **Q** is minimized by the Moore-Penrose generalized inverse as follows

$$\mathbf{Q} = \mathbf{L}^T(\mathbf{L}\mathbf{L}^T)^{-1} \cdot \mathbf{B} \quad (6)$$

where superscript T and -1 indicate the transpose and inverse of the matrix, respectively. The solution of MNE is found to be inappropriate for MEG source localization because it tends to give current distribution in shallower positions. The MNE explains the measured magnetic field with the least power of **Q**; therefore, the large deeper sources in actual current distribution are represented by weak shallower sources in the MNE. More appropriate solutions can be drawn in combination with normalization techniques, such as Tikhonov regularization and weighting, as shown below.

L2-norm minimization is to minimize the quadratic form of **Q** in Eq. (5), namely to minimize $\mathbf{Q}^T \mathbf{\Sigma}^{-1} \mathbf{Q}$ from all the **Q** that satisfies $\mathbf{B} = \mathbf{L} \cdot \mathbf{Q}$

$$\hat{\mathbf{Q}} = \arg \min_{Q} \{ \mathbf{Q}^T \mathbf{\Sigma}^{-1} \mathbf{Q} | \mathbf{B} = \mathbf{L} \cdot \mathbf{Q} \} \qquad (7)$$

where Σ is a positive definite symmetrical matrix of weighting factor for **Q**. It is determined by the relative distances between the mesh point and the sensor position to normalize the contribution of the unit current amplitude to the sensors. The solution of Eq. (7) is derived from Lagrange's method of undetermined multipliers as follows.

$$\hat{\mathbf{Q}} = \mathbf{\Sigma} \mathbf{L}^T (\mathbf{L} \mathbf{\Sigma} \mathbf{\Sigma}^T)^{-1} \cdot \mathbf{B} \qquad (8)$$

The current distributions obtained from L2-norm estimation tend to show extended active areas rather than the actual one [31]. Care should be taken to interpret the estimation results as it may look like a widely distributed source even when a highly localized source (e.g., primary sensory responses, epileptic spikes) is subjected to L2-norm estimation.

To overcome the inadequacy of pseudo-extended activation in an L2-norm estimation, source localization techniques using the L1-norm estimation have been introduced [32, 33]. $\hat{\mathbf{Q}}$ is determined to minimize the sum of absolute values of **Q**:

$$\hat{\mathbf{Q}} = \arg \min_{Q} \{ \mathbf{\Sigma} |\mathbf{Q}| | \mathbf{B} = \mathbf{L} \cdot \mathbf{Q} \} \qquad (9)$$

Such $\hat{\mathbf{Q}}$ is known to have sparse distribution; that is, the measured magnetic field is explained by the least number of current dipoles. Therefore, the L1-norm estimation gives a good interpretation when the actual neuronal activity is well localized. Conversely, the widely distributed neuronal sources may be underrepresented.

Spatial Filter

The spatial filter was originally developed during study of radar–sonar processing or an earthquake wave search. The time course of neuronal activation at an arbitrary point in the head conductor can be determined by the spatial filter, which is composed from a covariance matrix of the measure (MEG data). Other source localization methods introduced above use data in a single time window, but the spatial filter uses the time-series MEG data to estimate the signal space. It is assumed that the signal component is picked up by a group of sensors surrounding the neuronal source and therefore has temporal and spatial correlations. Meanwhile, the noise component is assumed to distribute randomly and therefore to be independent among sensors. Based on these assumptions, a virtual spatial filter is composed by weighting the temporally correlative components to pass through the neuronal activity only from a certain point in the brain. The spatial filter for each mesh point in the brain is thoroughly calculated and is applied to the measured magnetic field to reconstruct the neuronal current distribution. Currently, the adaptive beam-former [34] or synthetic aperture magnetometry (SAM) [35] is applied as a practical source localization tool in MEG. In short, the magnetic field at i th sensor in the time window of t is represented as $B_m^i(t)$. The magnetic field of all the sensors is indicated as the column vector of

$$\mathbf{B_m}(t) = [B_m^1(t), B_m^2(t), \ldots, B_m^N(t)]^T$$

and the temporal mean of covariance matrices in a certain time window is shown as $\mathbf{R} = \langle \mathbf{B_m}(t) \cdot (\mathbf{B_m}(t))^T \rangle$. The current sources are assumed to be current dipoles whose positions, orientations, and the moments are described as $\mathbf{r}$, $\eta(\mathbf{r}) = [\eta_x(\mathbf{r}), \eta_y(\mathbf{r}), \eta_z(\mathbf{r})]^T$, and $\mathbf{Q}(\mathbf{r},t)$ respectively. The sensitivity of the dipoles to the sensor at r with the orientation of $\eta(r)$ is represented, using a lead field $\mathbf{L}$, as $l(r) = \mathbf{L}(r) \cdot \eta(r)$. Here $\hat{\mathbf{Q}}(\mathbf{r},t)$, the estimation of $\mathbf{Q}(\mathbf{r},t)$, is given by a spatial filter as follows.

$$w(\mathbf{r}) = \frac{\mathbf{R}^{-1}l(\mathbf{r})}{l^T(\mathbf{r})\mathbf{R}^{-1}l(\mathbf{r})} \tag{10}$$

$$\hat{\mathbf{Q}}(\mathbf{r},t) = w^T(\mathbf{r})\mathbf{B_m}(t) \tag{11}$$

The spatial filter $w(\mathbf{r})$ gives a minimum square mean value of $\hat{\mathbf{Q}}(\mathbf{r},t)$ under the constraint condition of $w^T(\mathbf{r})l^T(\mathbf{r}) = 1$. The spatial filter is considered a promising visualization tool of higher brain activity because it can extract some large signal components without signal averaging [35] and reconstruct the neuronal activation at an arbitrary position in the brain. However, the general assumption for the spatial filter—the temporal independence of different neuronal sources—is not always applicable to practical MEG data. Modifications in spatial filter construction have been examined [36, 37] to overcome the signal cancellation problem in the correlated activations.

Conclusion

There are more than 70 MEG systems in Japan, mainly due to the clinical advantage of preoperative mapping of sensory areas or epileptic foci. More versatile and appropriate source localization techniques are required to further the application of MEG in the fields of neuroscience and cognitive science. Recently, several congregative signal processing tools, including averaging, filtering, signal processing, and source localization software [38, 39], have been developed to integrate the MEG data analysis procedures beyond the differences of MEG system manufacturers. The enrichment of this integrative software will further the utilization of MEG in combination with other promising brain imaging tools.

References

1. Friston KJ, Holmes AP, Worsley KJ, et al (1995) Statistical parametric maps in functional imaging: a general linear approach. Hum Brain Mapp 2:189–210.
2. Frackowiak RSJ, Friston KJ, Frith C, et al (2003) Human brain function (2nd ed). Academic, Amsterdam.
3. Kyuhou S, Okada YC (1993) Detection of magnetic evoked fields associated with synchronous population activities in the transverse CA1 slice of the guinea pig. J Neurophysiol 70(6):2665–2668.
4. Okada YC, Wu J, Kyuhou S (1997) Genesis of MEG signals in a mammalian CNS structure. Electroencephalogr Clin Neurophysiol 103(4):474–485.
5. Ilmoniemi RJ, Hämäläinen MS, Knuutila J (1985) The forward and inverse problems in the spherical model. In: Weinberg H, Stroink G, Katila T (eds) Biomagnetism: applications & theory. Pergamon, New York, pp 278–282.
6. Sarvas J (1987) Basic mathematical and electromagnetic concepts of the biomagnetic inverse problem. Phys Med Biol 32:11–22.
7. Ryhänen T, Seppä H, Ilmoniemi R, et al (1989) SQUID magnetometers for low-frequency application. J Low Temp Phys 76:287–386.
8. Hilgenfeld B, Strahmel E, Nowak H, et al (2003) Active magnetic shielding for biomagnetic measurement using spatial gradient fields. Physiol Meas 24(3):661–669.
9. Hämäläinen M, Hari R, Ilmoniemi RJ, et al (1993) Magnetoencephalography—theory, instrumentation, and application to noninvasive studies of the working human brain. Rev Modern Phys 65(2):413–497.
10. Tallon-Baudry C, Bertrand O, Pernier J (1999) A ring-shaped distribution of dipoles as a source model of induced gamma-band activity. Clin Neurophysiol 110(4):660–665.
11. Numminen J, Makela JP, Hari R (1996) Distributions and sources of magnetoencephalographic K-complexes. Electroencephalogr Clin Neurophysiol 99(6):544–555.
12. Sarvas J (1987) Basic mathematical and electromagnetic concepts of the biomagnetic inverse problem. Phys Med Biol 32(1):11–22.
13. Yang TT, Gallen CC, Schwartz BJ, et al (1993) Noninvasive somatosensory homunculus mapping in humans by using a large-array biomagnetometer. Proc Natl Acad Sci USA 90(7):3098–3102.
14. Nakamura A, Yamada T, Goto A, et al (1998) Somatosensory homunculus as drawn by MEG. Neuroimage 4(Pt 1):377–386.
15. Nakasato N, Seki K, Fujita S, et al (1996) Clinical application of visual evoked fields using an MRI-linked whole head MEG system. Front Med Biol Eng 7(4):275–283.
16. Supek S, Aine CJ, Ranken D, et al (1999) Single vs. paired visual stimulation: superposition of early neuromagnetic responses and retinotopy in extrastriate cortex in humans. Brain Res 830(1):43–55.
17. Schiffbauer H, Berger MS, Ferrari P, et al (2003) Preoperative magnetic source imaging for brain tumor surgery: a quantitative comparison with intraoperative sensory and motor mapping. Neurosurg Focus 15(1):E7.
18. Cappell J, Schevon C, Emerson RG (2006) Magnetoencephalography in epilepsy: tailoring interpretation and making inferences. Curr Neurol Neurosci Rep 6(4):327–331.
19. Knowlton RC, Shih J (2004) Magnetoencephalography in epilepsy. Epilepsia 45(suppl):61–71.
20. Nakasato N, Levesque MF, Barth DS, et al (1994) Comparisons of MEG, EEG, and ECoG source localization in neocortical partial epilepsy in humans. Electroencephalogr Clin Neurophysiol 91(3):171–178.
21. Kakigi R, Hoshiyama M, Shimojo M, et al (2000) The somatosensory evoked magnetic fields. Prog Neurobiol 61(5):495–523.
22. Kakigi R, Inui K, Tamura Y (2005) Electrophysiological studies on human pain perception. Clin Neurophysiol 116(4):743–763.

23. Chen R, Hallett M (1999) The time course of changes in motor cortex excitability associated with voluntary movement. Can J Neurol Sci 26(3):163–169.
24. Weinberg H, Cheyne D, Crisp D (1990) Electroencephalographic and magnetoencephalographic studies of motor function. Adv Neurol. 54:193–205.
25. Jacobson GP (1994) Magnetoencephalographic studies of auditory system function. J Clin Neurophysiol 11(3):343–364.
26. Mizoguchi C, Kobayakawa T, Saito S, et al (2002) Gustatory evoked cortical activity in humans studied by simultaneous EEG and MEG recording. Chem Senses 27(7):629–634.
27. Tonoike M, Yamaguchi M, Kaetsu I, et al (1998) Ipsilateral dominance of human olfactory activated centers estimated from event-related magnetic fields measured by 122-channel whole-head neuromagnetometer using odorant stimuli synchronized with respirations. Ann NY Acad Sci 855:579–590.
28. Mosher JC, Lewis PS, Leahy RM (1992) Multiple dipole modeling and localization from spatio-temporal MEG data. IEEE Trans Biomed Eng 39(6):541–557.
29. Lin FH, Witzel T, Hamalainen MS, et al (2004) Spectral spatiotemporal imaging of cortical oscillations and interactions in the human brain. Neuroimage 23(2):582–595.
30. Fujimaki N, Hayakawa T, Nielsen M, et al (2002) An fMRI-constrained MEG source analysis with procedures for dividing and grouping activation. Neuroimage 17(1):324–343.
31. Lantz G, Spinelli L, Menendez RG, et al (2001) Localization of distributed sources and comparison with functional MRI. Epileptic Disord Special Issue:45–58.
32. Matsuura K, Okabe Y (1995) Selective minimum-norm solution of the biomagnetic inverse problem. IEEE Trans Biomed Eng 42(6):608–615.
33. Uutela K, Hämäläinen M, Somersalo E (1999) Visualization of magnetoencephalographic data using minimum current estimates. Neuroimage 10:173–180.
34. Sekihara K, Nagarajan SS, Poeppel D, et al (2001) Reconstructing spatio-temporal activities of neural sources using an MEG vector beamformer technique. IEEE Trans Biomed Eng 48(7):760–771.
35. McCubbin J, Vrba J, Spear P, et al (2004) Advanced electronics for the CTF MEG system. Neurol Clin Neurophysiol 69 (online journal).
36. Sekihara K, Nagarajan SS, Poeppel D, et al (2002) Performance of an MEG adaptive-beamformer technique in the presence of correlated neural activities: effects on signal intensity and time-course estimates. IEEE Trans Biomed Eng 49(Pt 2):1534–1546.
37. Kimura T, Kako M, Kamiyama H, et al (2007) Inverse solution for time-correlated multiple sources using beamformer method. International Congress Series 1300 (in press).
38. FieldTrip software. Available from F.C. Donders Centre for Cognitive Neuroimaging (http://www.ru.nl/fcdonders/fieldtrip/).
39. Dalal SS, Zumer JM, Agrawal V, et al (2004) NUTMEG: a neuromagnetic source reconstruction toolbox. Neurol Clin Neurophysiol 52 (online journal).

Part II
Related Topics

Section I
Learning and Memory

Interactions Between Chewing and Brain Activity in Humans

M. Onozuka[1,2], Y. Hirano[1,3], A. Tachibana[1,2], W. Kim[1,2], Y. Ono[1,3], K. Sasaguri[2,4], K. Kubo[5], M. Niwa[6], K. Kanematsu[7], and K. Watanabe[8]

Summary

The involvement of chewing in brain activity in humans has been studied. In our studies using functional magnetic resonance imaging (fMRI) and behavioral techniques, chewing resulted in a bilateral increase in blood oxygenation level-dependent (BOLD) signals in the sensorimotor cortex, supplementary motor area, insula, thalamus, and cerebellum. In addition, in the first three regions, chewing moderately hard gum produced stronger signals than chewing hard gum. However, in the aged group, the BOLD signal increases were smaller in the first three regions and higher in the cerebellum. Only the aged subjects showed significant increases in various association areas to which input activities in the primary sensorimotor cortex, supplementary area, or insula had positive path coefficients. Furthermore, chewing ameliorates the age-related decrease in hippocampal activities during encoding and that in retrieval memory. The findings suggest the involvement of chewing in memory processes.

---

[1]Departments of Physiology and Neuroscience, Kanagawa Dental College, 82 Inaoka-cho, Yokosuka 238-8580, Japan

[2]Research Center of Brain and Oral Science, Kanagawa Dental College, 82 Inaoka-cho, Yokosuka 238-8580, Japan

[3]Department of Molecular Imaging Center, National Institute of Radiological Sciences, 4-9-1 Anagawa, Inage-ku, Chiba 263-8555, Japan

[4]Department of Orthodontics, Kanagawa Dental College, 82 Inaoka-cho, Yokosuka 238-8580, Japan

[5]Department of Oral Anatomy, Asahi University School of Dentistry, 1851 Hozumi, Mizuho 501-0296, Japan

[6]Department of Radiology, Tono Welfare Hospital, 76-1 Toki, Mizunami 509-6101, Japan

[7]Department of Radiology, Gifu University Graduate School of Medicine, 1-1 Yanagido, Gifu 501-1194, Japan

[8]Physiological Laboratory, Faculty of Care and Rehabilitation, Seijoh University, 2-172 Fukinodai, Tokai 476-8588, Japan

Key words Chewing, Learning and memory, Brain activation, Hippocampus, Prefrontal area, Aging

Introduction

Chewing is the process by which food is mashed and crushed by teeth. It is the first step of digestion and increases the surface area of foods to allow more efficient breakdown by enzymes. During the mastication process, the food is positioned between the teeth for grinding by the cheek and tongue. As chewing continues, the food is made softer and warmer, and the enzymes in saliva begin to break down carbohydrates in the food. After chewing, the food (now called a bolus) is swallowed. It enters the esophagus and continues on to the stomach, where the next step of digestion occurs.

In this digestion process, chewing is a complex activity requiring orchestration of the nuclear groups supplying the muscles that move the mandible, tongue, cheeks, and hyoid bone. The chief controlling center seems to be an area of the premotor cortex directly in front of the face representation on the motor cortex. Stimulation of this area produces chewing cycles [1]. In contrast, the lower part of the trigeminothalamic tract commences in the spinal trigeminal nucleus [1, 2]. Nearly all of these fibers cross the midline before ascending into the pons. This component has features in common with the spinal lemniscus, which accompanies it in the brain stem, mediating tactile, nociceptive, and thermal sensations [2, 3]. In the pons, it is jointed by fibers crossing from the principal sensory nucleus, thus completing the trigeminal lemniscus, which terminates in the ventral posterior medial nucleus of the thalamus. From the thalamus, third-order afferents project to the large area of facial representation in the lower half of the somatic sensory cortex.

In recent years, several investigators have pointed out that chewing plays an important role in the promotion and preservation of general health. For example, chewing has been shown to ameliorate reduced nutritional status, overall health, and activities of daily living in the elderly [4]. According to the World Health Organization/International Dental Federation Goals for the Year 2000, reducing the number of missing teeth in the elderly is a primary aim. Also, in Japan, the "8020" program (at the age of 80, at least 20 remaining teeth) is gaining popularity for the maintenance of sufficient chewing performance throughout a lifetime. Furthermore, it has been suggested that masticatory dysfunction is related to the development of senile dementia. A direct relation between the progress of dementia and a reduced ability to masticate owing to a decreased number of residual teeth, decreased use of dentures, or decreased maximal biting force has been reported in elderly humans with dementia [2], although the mechanism(s) underlying the relation between masticatory dysfunction and impaired learning and memory is not well understood. In this chapter, we report our findings obtained with functional magnetic resonance imaging (fMRI) and behavioral techniques, which suggest the interaction between chewing and brain activation in humans.

Brain Regional Activation by Chewing

It has been shown that in humans gum chewing not only results in transient increases in energy expenditure and heart rate response [5], it increases cerebral blood flow due to changes in internal carotid arterial blood flow [6, 7]. Furthermore, cerebral blood flow imaging during gum chewing, revealed by positron emission tomography (PET), shows increased blood flow in the bilateral lower frontal and parietal lobes [8, 9]. However, because of the low spatial and temporal resolution of PET, it is difficult to record actual brain activation during chewing and to identify the fine anatomical regions activated. To test specific hypotheses about the anatomical and physiological regions involved in processing sensory and motor information in the human brain [10], functional magnetic resonance imaging (fMRI), because of its enhanced spatial and temporal resolution [11], offers the advantage that both the actual response to chewing and the fine regions linked to chewing can be analyzed.

In our experiment using fMRI to evaluate brain activation associated with chewing in intact humans [12], the task paradigm was rhythmic chewing, at a rate of approximately 1 Hz, of two kinds of gum: moderately hard gum (X type, 5.6×10^4 poise) and hard gum (G type, 2.3×10^5 poise) [13]. These gums—essentially chewing gum without the odor and taste components—were specially prepared in the General Laboratory of Lotte Co. (Saitama, Japan). Each subject performed four cycles of 32 s of rhythmic chewing and 32 s without chewing (Fig. 1A).

In all subjects (ages 20–31 years), gum chewing was associated with significant increases in the blood oxygenation level-dependent (BOLD) signal in various regions of the brain (Figs. 1B,C). An increase was seen bilaterally in the primary sensorimotor cortex extending down into the upper bank of the operculum and insula. In addition, increases were seen in the supplementary motor area (extending down into the cingulate gyrus), thalamus, and cerebellum. Figure 2 shows the chewing-related signal changes in these regions. The locations of the most significant foci of activation for these regions are summarized in Table 1, in which the anatomical regions with maximum *t*-values in clusters with Talairach coordinates are shown. In this experiments, during gum chewing the strongest activation was found in the primary sensorimotor cortex. Penfield and Boldrey [14], who mapped the primary motor cortex in humans by electrically stimulating the exposed scalp during neurosurgery, demonstrated that the masticatory organs are represented on the inferior aspect of the primary motor cortex, close to the lateral fissure. Their mapping is in good agreement with the increased BOLD signal seen in the primary motor cortex during gum chewing. We also showed relatively large bilateral regions of activation that encompassed the inferolateral primary sensory cortex, in which the masticatory organs are present [14].

Studies in nonhuman primates have shown that many "nonprimary" motor areas in the cortex are associated with voluntary control of movement [15]. Some of these areas show somatotopic mapping and have direct connections to the primary motor cortex or spinal cord. Fink et al. [15] used PET to map the location of such areas

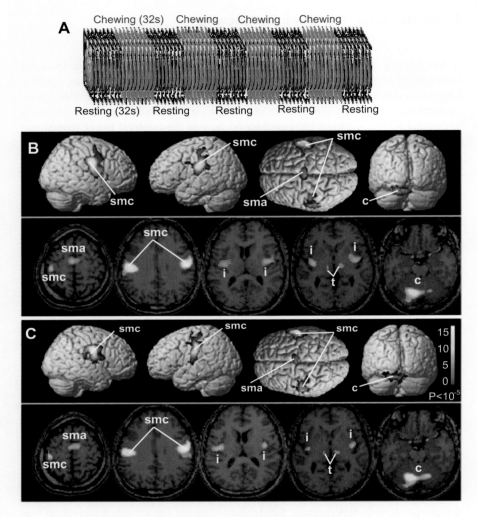

Fig. 1. Brain regional activities. **A** Task paradigm. **B, C** Significant blood oxygenation level-dependent (BOLD) functional magnetic resonance imaging (fMRI) signal increases associated with chewing X (moderately hard) gum (**B**) and G (hard) gum (**C**) by individual subjects. *Upper section*: activated areas superimposed on a template. *Lower section*: activated regions superimposed on a T1-weighted MRI scan. *smc*, primary sensorimotor cortex; *sma*, supplementary motor area; *i*, insula; *t*, thalamus; *c*, cerebellum. Color scale: *t* value. Group analysis was used

in humans using voluntary movements of the hand, shoulder, and leg. It is notable that all the nonprimary cortical areas associated with chewing identified in the present study (i.e., the supplementary motor area and the insula), also called the masticatory center [2], were also identified during hand, shoulder, and leg movements [15]. Additionally in our study, chewing was associated with significant signal increases in the thalamus and cerebellum, typical of those associated

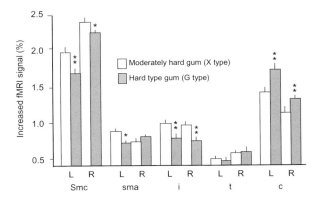

Fig. 2. Comparison of the increased fMRI signals (%) obtained using X (*open column*) or G (*filled column*) type gum. L, left side; R, right side. Each column represents the mean ± SE (n = 14). *P < 0.05 compared to X type gum. **P < 0.01 compared to X-type gum

Table 1. Locations of the most significant foci of activation for various regions

Region of activation	BA	Maximal t-value	Talairach coordinates of max-voxel		
			x	y	z
X-type gum					
L sensorimotor cortex	3, 4	15.43	−48	−18	32
R sensorimotor cortex	3, 4	16.89	52	−14	30
L supplementary motor area	6	8.26	−4	−8	56
R supplementary motor area	6	6.01	8	−8	64
L insula	13	10.58	−40	−12	16
R insula	13	8.48	42	−6	12
L thalamus		5.34	−12	−20	0
R thalamus		5.64	14	−18	2
L cerebellum		12.67	−8	−68	−22
R cerebellum		8.33	20	−66	−28
G-type gum					
L sensorimotor cortex	3, 4	13.29	−48	−20	38
R sensorimotor cortex	3, 4	15.06	56	−12	30
L supplementary motor area	6	6.97	−2	−6	60
R supplementary motor area	6	7.00	10	−6	58
L insula	13	8.03	−42	−14	16
R insula	13	7.81	40	−8	14
L thalamus		4.98	−10	−20	0
R thalamus		6.12	14	−20	2
L cerebellum		12.26	−14	−66	−22
R cerebellum		9.47	18	−66	−26

X-type gum, moderately hard gum; G-type gum, hard gum; R, right; L left

with voluntary control of movement [16]. A significant increase in the BOLD signal was seen throughout the striatum, prefrontal cortex, or parietal cortex (Fig. 2), in which blood flow has been shown to be increased by gum chewing [8]. However, the peak of the BOLD signal in the striatum could not be isolated owing to strong activation of the insula, and the locations of the signal changes in the prefrontal or parietal cortex were different in each subject. Thus, we cannot rule out the possibility that these signal changes might contain some artifacts or represent unknown mechanisms.

In the cerebellum, chewing hard gum caused a greater increase in the BOLD signal than did chewing moderately hard gum. It has been shown that an increase in the masticatory force elevates activity in the masseter muscle [17], where sensory information is finally projected to the cerebellum [18]. Thus, it is reasonable to state that a greater increase in the signal when chewing hard gum reflects increased information from the masticatory muscle.

Our results showing lower cortical activation when chewing hard gum compared to moderately hard gum are in good agreement with the findings that mastication of moderately hard food leads to a greater increase in cerebral blood flow than that of either soft or hard food [3, 7]. Together with the fact that cerebral activation while chewing a gelatin dessert is low, it suggests that chewing with a moderate biting force may be most effective in maintaining neuronal activity in the brain.

Age-Dependent Attenuation of Brain Activation by Chewing

We also used fMRI to assess the effect of aging on brain regional activity associated with chewing, as there was a relation between oral status and the progress of dementia [19]. Three groups of neurologically healthy subjects were included in this study: a young adult group (ages 19–26 years), a middle-aged group (ages 42–55 years), and an aged group (ages 65–73 years). It was found that in the primary sensorimotor cortex, cerebellum, and thalamus the chewing-induced increase in the BOLD signal was attenuated in an age-dependent manner (Fig. 3). A study on aging and mastication have shown that the loss of teeth and the masticatory muscle power deficits seen with advancing age impair masticatory function, thereby causing a reduction in sensory input activity to the central nervous system [20]. In the experiments, biting force was highest in the young adult group, followed by the middle-aged group, and then the aged group. A similar age-dependent decline was seen in the number of remaining teeth. Taken together with the fact that age-related degeneration of various brain regions, including the somatosensory cortex, occurs in humans [21], it is suggested that the age-related attenuation of the signal seen in the above three regions results from an age-dependent decrease in both masticatory work and neuronal activity in the brain.

However, when an individual analysis method was employed, we found important evidence (Fig. 4) showing that signal increases in a variety of association areas

Chewing and Brain Activation

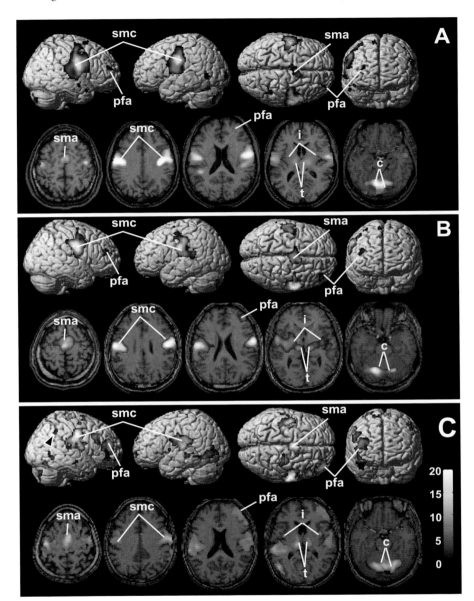

Fig. 3. Effect of aging on brain regional activity during chewing. Significant signal increases associated with gum chewing in a young adult subject (**A**), a middle-aged subject (**B**), and an aged subject (**C**). *Upper section*: activated areas superimposed on a template ($P < 0.05$, corrected for multiple comparisons). *Lower section*: activated regions superimposed on a T1-weighted MRI scan ($P < 0.001$, uncorrected for multiple comparisons). *pfa*, prefrontal area. Color scale: *t* value (degrees of freedom = 87.12)

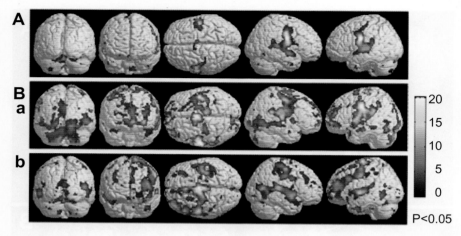

Fig. 4. Brain regional activities. Significant signal increases associated with chewing moderately hard gum by an individual young subject (**A**, 21-year-old woman) and aged subjects (**B,a**, 69-year-old woman; **B,b**, 73-year-old man). The individual analysis was used. Note the different activation patterns in the association areas between the two aged subjects

were noted in aged subjects in addition to the regions having significant signal increases during the group analysis (compare Fig. 3). Furthermore, the locations of activation foci in these areas were different in each subject. Such phenomena were never seen in the young subjects.

Chewing and Prefrontal Cortex

To evaluate the neuronal mechanism underlying chewing-induced activation in the association areas in the aged subject, we examined the psychophysiological interaction with a structural equation model using AMOS software [22]. As a result, it was found that gum chewing-related input activities in the association areas from the sensorimotor cortex, supplementary motor area, and insula are transported via corticocortical pathways to the prefrontal association cortex (Fig. 5).

Surprisingly, our findings indicate that, in all groups, gum chewing resulted in an increased BOLD signal in the right prefrontal area; and this increase was four times higher in aged subjects than in young subjects. A previous PET study found that during cognitive tasks patients with early Alzheimer's disease show increased activity in the prefrontal regions compared to that in healthy age-matched controls [23]. Furthermore, these authors showed that increased right prefrontal cortex activity is associated with better memory performance in both groups [23];

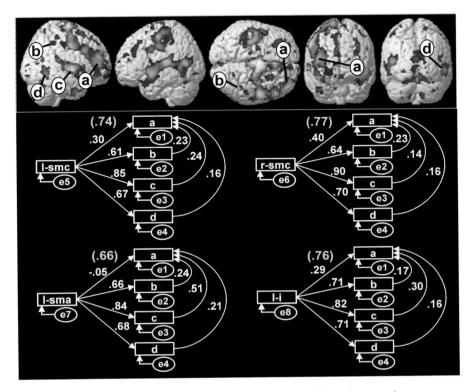

Fig. 5. Path diagram using structural equation modeling involving the sensorimotor cortex, supplementary motor area, insula, and association areas (a, b, c, d, respectively). Each value indicates the standardized path coefficients under the condition of gum chewing. e, error variable

this has been interpreted as compensatory reallocation of cognitive resources [24].

All regions of the human prefrontal cortex are interconnected. This cortex is also richly interconnected with other areas of the brain. For example, the prefrontal cortex has at least five distinct regions, each of which is independently involved in separate corticostriatal loops [25]. The prefrontal cortex also has strong limbic system connections via its medial and orbital efferent connections that terminate in the amygdala, thalamus, and parahippocampal regions [25]. Finally, it has long pathway connections to association cortex in the temporal, parietal, and occipital lobes. Almost all of these pathways are reciprocal. Thus, if the interpretation of Grady et al. [23, 24, 26] is correct, it is possible that in the elderly chewing stimulates neuronal activity in a network between the right prefrontal cortex and the hippocampus, which might be useful in maintaining cognitive function.

Chewing and Hippocampal Activities

In general, the hippocampus is the first region where new memory is formed. This region is located in the central part of the cerebrum and receives all sensory information. For example, according to recent evidence, visual information first enters the visual cortex and then enters the hippocampus, where it is determined whether the information is useful [27, 28]. Only useful information is transported to various cerebral regions and stored there for memorizing [27].

Therefore, together with the fact that chewing activates a variety of association areas in which sensory information is projected via the perforant path into the hippocampus, a role of increased neuronal activities in the association area during chewing in the elderly may be hypothesized. If this hypothesis is true, neuronal activities in the hippocampus of the elderly must be increased by chewing. In studies using humans and animals, many investigators have shown that spatial cognitive function is localized in the hippocampus.

Thus, we did another fMRI experiment in which the effect of chewing on the hippocampal activities was examined using a task for spatial cognition. As shown in Fig. 6A, subjects were shown 16 photographs followed by the same number of pictures of a plus character (+) on a green background during each cycle. Each picture or photograph was projected every 2 s during the cycle. subjects were asked to remember as many of the photographs as they could. During this photograph encoding, the BOLD signal in the hippocampus was continuously recorded. Figure 6B is an example of the result from a 21-year-old woman in whom the hippocampus was strongly activated but no significant difference was seen before and after chewing. In contrast, activation in the elderly was quite small in comparison with that in the young subject. However, the activation area and the intensity of the fMRI signals were increased by chewing. Figure 7A shows BOLD signal changes in the hippocampus of a 74-year-old woman before chewing, indicating that the BOLD signals are linked with the encoding task (upper trace, Fig. 7B). Furthermore, when the mean value in the resting condition was subtracted from the mean value obtained for the encoding condition, an increased value of 0.2% was estimated in the case of no chewing. In contrast, after chewing, we found that the signal intensity doubled, that is, it was 0.4% (lower traces in Figs. 7A and 7B).

Based on these results, chewing seems to induce an increase in sensory input from receptors linked to it, and chewing-associated face and jaw movements increase the sensory input from the somatic sensory cortex, which is involved in circuits in the hippocampus. With respect to Alzheimer's disease and aging, the single most vulnerable circuit in the cerebral cortex is the projection referred to as the perforant path [29], which originates in the entorhinal cortex and terminates in the dentate gyrus, thus providing the key interconnection between the neocortex and hippocampus [27]. The entorhinal cortex is a region of extraordinary convergence of inputs from the association cortex, essentially funneling highly processed neocortical information into the dentate gyrus of the hippocampus and playing a crucial role in learning and memory [30].

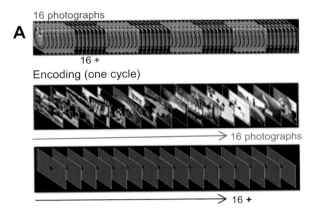

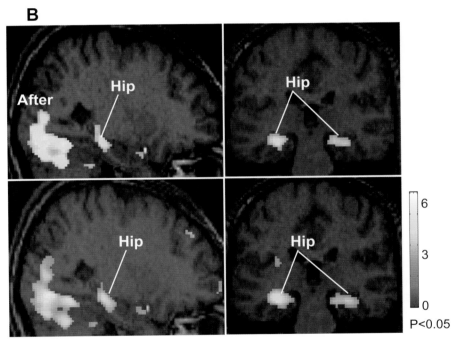

Fig. 6. Hippocampal activity in a young subject. **A** Task paradigm. **B** Significant signal increases associated with photograph encoding before and after gum chewing. *Hip*, hippocampus. Color scale: t value

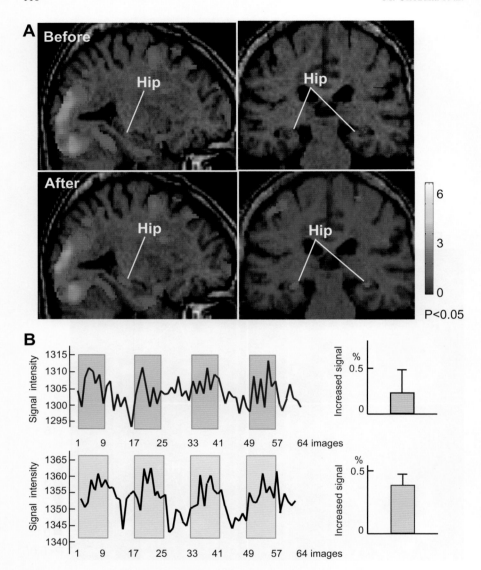

Fig. 7. Hippocampal activities in an aged subject. **A** Significant signal increases are associated with photograph encoding before and after gum chewing. Color scale: *t* value. **B** Changes in signal intensity on an image-by image basis for 64 successive images during four cycles of encoding of photographs: brown (without chewing) and pink (with chewing) boxes; plus (+) characters (without boxes)

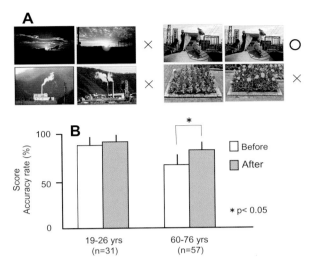

Fig. 8. Memory acquisition before and after gum chewing for 2 min. A recall test was carried out 20 min after the encoding experiments. In the recall test, we used 64 photographs at random: 32 of the photographs were repeated from the encoding test, and the other 32 photographs were newly added. The subject had to judge whether each photograph had been seen before

Chewing and Memory Acquisition

We tested whether a chewing-induced increase in hippocampal activities is reflected in an increase in learning ability and memory acquisition. Twenty minutes after encoding experiments, a recall test was carried out.

In the recall test, we used 64 photographs at random: 32 of the photographs were repeated from the encoding test, and the other 32 photographs were newly added (Fig. 8A). The subject had to judge whether each photograph had been seen before. Figure 8B shows the results of the retrieval test. Aged subjects showed a significant increase in memory acquisition after chewing for 2 min, whereas chewing had no effect in young subjects. These findings suggest the link between an increase in hippocampal BOLD signals and enhanced memory acquisition.

Conclusion

Based on the above findings, we propose that chewing by the elderly indirectly activates the association areas by mediating the neuronal circuits in the mastication center. Sensory information in this center is projected through the perforant path into the hippocampus. The hippocampus is the first brain region to show neuropathological changes with advancing age and plays a crucial role in learning and memory. Furthermore, chewing warms up neuronal activities in the circuit between the hippocampus and the prefrontal area, which plays an important role in maintaining input activities in the brain that are indispensable for suppressing neuronal degeneration. Therefore, we strongly suggest that chewing is a useful therapy for preventing senile dementia.

Acknowledgments Our studies were supported by Grants-in-Aid for Scientific Research from the Ministry of Education, Science, and Culture of Japan and a grant from Lotte Co. Ltd. (Tokyo, Japan).

References

1. FitzGerald MJT, Folan-Curran J (2002) Clinical neuroanatomy and related neuroscience. Saunders, London, pp 178–180.
2. Nakata M (1998) Masticatory function and its effects on general health. Int Dent J 48: 540–548.
3. Nakamura Y, Katakura N (1995) Generation of masticatory rhythm in the brainstem. Neurosci Res 23:1–19.
4. Miura H, Araki Y, Hirai T, et al (1998) Evaluation of chewing activity in the elderly person. J Oral Rehabil 25:190–193.
5. Suzuki M, Shibata M, Sato Y (1992) Energy metabolism and endocrine responses to gum-chewing. J Mastica Health Sci 2:55–62.
6. Sasaki A (2001) Influence of mastication on the amount of hemoglobin in human brain tissue. Kokubyo Gakkai Zasshi 68:72–81.
7. Sesay M, Tanaka A, Ueno Y, et al (2000) Assessment of regional cerebral blood flow by xenon-enhanced computed tomography during mastication in humans. Keio J Med 49(suppl 1):A125–A128.
8. Momose I, Nishikawa J, Watanabe T, et al (1997) Effect of mastication on regional cerebral blood flow in humans examined by positron-emission tomography with ^{15}O-labelled water and magnetic resonance imaging. Arch Oral Biol 42:57–61.
9. Watanabe I, Ishiyama N, Senda M (1992) Cerebral blood flow during mastication measures with positron emission tomography. Geriatr Dent 6:148–150.
10. Yancey SW, Phelps EA (2001) Functional neuroimaging and episodic memory: a perspective. J Clin Exp Neuropsychol 23:32–48.
11. Meisenzahl EM, Schlosser R (2001) Functional magnetic resonance imaging research in psychiatry. Neuroimaging. Clin N Am 11:365–374.
12. Onozuka M, Fujita M, Watanabe K, et al (2002) Mapping brain region activity during chewing: a functional magnetic resonance imaging study. J Dent Res 81:743–746.
13. Suzuki M, Ishiyama I, Takiguchi T, et al (1994) Effects of gum hardness on the response of common carotid blood flow volume, oxygen uptake, heart rate and blood pressure to gum-chewing. J Mastica Health Sci 4:9–20.
14. Penfield W, Boldrey E (1938) Somatic motor and sensory representation in the cerebral cortex as studied by electrical stimulation. Brain 15:389–443.
15. Fink GR, Frackowiak RSJ, Pietrzyk U, et al (1977) Multiple non-primary motor areas in the human cortex. J Neurophysiol 77:2164–2174.
16. Passingham RE (1993) The frontal lobes and voluntary action. Oxford University Press, Oxford.

17. Proschel PA, Raum J (2001) Preconditions for estimation of masticatory forces from dynamic EMG and isometric bite force-activity relations of elevator muscles. Int J Prosthodont 14:563–569.
18. Kubota K, Nagae K, Shibanai S, et al (1988) Degenerative changes of primary neurons following tooth extraction. Anat Anz 166:133–139.
19. Okimoto K, Ieiri K, Matsuo K, et al (1991) Ageing and mastication: the relationship between oral status and the progress of dementia at senile hospital. J Jpn Prosthodont Soc 35: 931–943.
20. Onozuka M, Fujita M, Watanabe K, et al (2003) Age-related changes in brain regional activity during chewing: a functional magnetic resonance imaging study. J Dent Res 82:657–660.
21. Godde B, Berkefeld T, David-Jurgens M, et al (2002) Age-related changes in primary somatosensory cortex of rats: evidence for parallel degenerative and plastic-adaptive processes. Neurosci Biobehav Rev 26:743–752.
22. Hirano Y, Fujita M, Watanabe K, et al (2006) Effect of unpleasant loud noise on hippocampal activities during picture encoding: an fMRI study. Brain Cogn 61:280–285.
23. Grady CL, Furey ML, Pietrini P, et al (2001) Altered brain functional connectivity and impaired short-term memory in Alzheimer's disease. Brain 124:739–756.
24. Grady CL, McIntosh AR, Beig S, et al (2001) An examination of the effects of stimulus type, encoding task, and functional connectivity on the role of right prefrontal cortex in recognition memory. Neuroimage 14:556–571.
25. Boller F, Grafman J (eds) (2000) Handbook of neuropsychology (Vol 7, 2nd ed). Elsevier, Amsterdam, pp 157–174.
26. Grady CL, McIntosh AR, Beig S, et al (2003) Evidence from functional neuroimaging of a compensatory prefrontal network in Alzheimer's disease. J Neurosci 23:986–993.
27. Amaral DG, Witter MP (1989) The three dimensional organization of the hippocampal formation: a review of anatomical data. Neuroscience 31:571–591.
28. FitzGerald MJT, Folan-Curran J (2002) Handbook of neuropsychology (Vol 2, 2nd ed)., Elsevier, Tokyo, pp 49–65.
29. Squire LR, Zola-Morgan S (1991) The medial temporal lobe memory system. Science 253: 1380–1386.
30. Wainer BH, Steininger TL, Roback JD, et al (1993) Ascending cholinergic pathways: functional organization and implications for disease models. Prog Brain Res 98:9–30.

Involvement of Dysfunctional Mastication in Cognitive System Deficits in the Mouse

Kazuko Watanabe[1], Kin-ya Kubo[2], Hiroyuki Nakamura[3], Atsumichi Tachibana[4,5], Wanjae Kim[5], Yumie Ono[4,5], Kenichi Sasaguri[5,6], and Minoru Onozuka[4,5]

Summary

A systemic effect of dysfunctional mastication has been suggested as a possible epidemiological risk factor for senile dementia. In recent years, we have evaluated the effects on cognitive function deficits in SAMP8 mice. Aged mice with dysfunctional mastication showed significantly reduced learning ability in a water maze test compared with age-matched control mice, whereas there was no difference between control and molarless young adult mice. Immunohistochemical analysis revealed that in the CA1 region of the hippocampus the molarless condition not only enhanced the age-dependent increase in the density and hypertrophy of GFAP-labeled astrocytes, it decreased the density of Fos-positive neurons and Nissl-stained neurons, or the amount of acetylcholine (ACh) release in the hippocampus, in the same manner. There was a similar age-dependent decrease in choline acetyltransferase in the medial septal nucleus. Furthermore, dysfunctional mastication induced an increase in plasma corticosterone levels. The findings suggest that dysfunctional mastication in aged SAMP8 mice causes abnormalities in the hippocampus through stress, leading to deficits in learning and memory.

Key words Mastication, Hippocampus, Memory, Neuronal degeneration, Stress

---

[1]Faculty of Care and Rehabilitation (Physiology), Seijoh University, Fukinodai 2-172, Tokai, Aichi 476-8588, Japan

[2]Department of Oral Anatomy, Asahi University School of Dentistry, 1851 Hozumi, Mizuho 501-0296, Japan

[3]Department of Morphological Neuroscience, Gifu University Graduate School of Medicine, 1-1 Yanagido, Gifu 501-1194, Japan

[4]Department of Physiology and Neuroscience, Kanagawa Dental College, 82 Inaoka-cho, Yokosuka 238-8580, Japan

[5]Research Center of Brain and Oral Science, Kanagawa Dental College, 82 Inaoka-cho, Yokosuka 238-8580, Japan

[6]Department of Orthodontics, Kanagawa Dental College, 82 Inaoka-cho, Yokosuka 238-8580, Japan

Introduction

A reduced ability to masticate as a result of root caries, a large number of missing teeth, and oral dyskinesia have been suggested to be associated with senile dementia. For example, a systemic effect of tooth loss has been suggested as a possible epidemiological risk factor for Alzheimer's dementia [1]. In experimental studies, it has also been shown that a soft diet from the weanling period onward causes later impairment of avoidance performance in mice [2], and that differences in neuronal density between the right and left cerebral hemispheres are seen in rats with unilateral mastication [3]. Recently, many investigators have extensively studied the mechanism(s) connecting dysfunctional mastication with senile deficits in cognition.

In this chapter, we describe our recent findings obtained using behavioral, immunohistochemical, and biochemical methods in the hippocampus of mice with accelerated senescence (SAMP8 mice).

Impaired Spatial Cognition in the Molarless Condition

To achieve an animal model with dysfunctional mastication, we extracted or cut off the maxillary molar teeth (molarless condition) of SAMP8 mice. Control animals underwent the same surgical procedure except for tooth removal or extraction. Ten days after operating, animals were tested in the spatial memory version of the Morris water maze [4]. In this behavioral experiment, the mice were placed in a pool of nontranslucent water and given spatial cues to help them find a hidden platform [5]. Figure 1 shows the time required for the mice to arrive at the platform (escape latency), revealing that the escape latency of the aged group was longer than that of the young group. In control mice, significant overall differences were seen between the three age groups in terms of the time taken to reach the platform

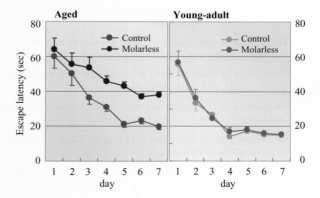

Fig. 1. Spatial learning in the water maze test. Ten days after surgery, learning tests were started. The results are expressed as the mean score of four trials per day. Note that molarless mice take a significant longer time to reach the platform than do controls

[5], indicating age-related impairment of spatial learning. Although young molarless mice performed as well as their age-matched controls, middle-aged and aged molarless mice performed significantly less well than their age-matched controls. When a visible probe test was performed on the same mice at the end of the test, no significant difference was seen between the control and molarless groups [5–8]. In addition, the swimming speed of each mouse, estimated from the latency and swim paths recorded in the computer system, showed no difference between the age-matched control and molarless groups.

At 6 months of age, SAMP8 mice exhibit clear deficits in learning and memory in various learning tests (e.g., passive avoidance test [9, 10] and one-way [9], T maze [11], and Sidman active avoidance tests [12]), suggesting that the molarless condition enhances the age-dependent decrease in cognitive function. Kubota et al. [13] demonstrated that sensory input from the sensory receptor on the tooth root decreases as a result of long-term dysfunction of mastication following extraction of teeth. Kato et al. [14] also showed impaired performance in a radial arm maze task in aged rats lacking their molar teeth, implying that the molarless condition is linked to age-related memory deficits. Several studies have also shown that mastication increases muscular activity, regardless of caloric intake, and that brain neuronal activity and cerebral blood flow increase simultaneously [15, 16]. Thus, it is likely that there may be an important link between mastication-induced stimulation and hippocampal function.

Attenuation of Hippocampal Fos Induction by the Molarless Condition

In our study using the combined technique of behavior and immunohistochemistry in aged SAMP8 mice, the molarless condition decreased Fos induction in the hippocampus, which is related to impaired development of learning ability in a water maze [6]; this suggests that functional molar teeth may be one of the factors responsible for maintaining spatial learning and memory during old age [16]. This conclusion derives from the following findings: (1) in molarless mice, the reduction in Fos-positive cell numbers in the hippocampal CA1 region paralleled the reduction in learning ability in a water maze [6, 8]; and (2) the extent of the decrease in the number of these cells and the reduced learning ability [5, 6, 17] both depended on the duration of the molarless condition. Furthermore, as shown in Fig. 2, it was found that the suppressive effects induced by the molarless condition were considerably reduced by restoring the lost molar with an artificial crown [6].

The mechanism by which dysfunctional molar teeth and the consequent reduction in masticatory ability accelerate senile impairment of spatial learning has not yet been defined. However, one possibility is that in the molarless condition input activity to the somatic sensory cortex (in which some neurons are indirectly linked with hippocampal neurons) from receptors coupled to mastication and mastication-associated face and jaw movements may decrease. In addition, given the fact that

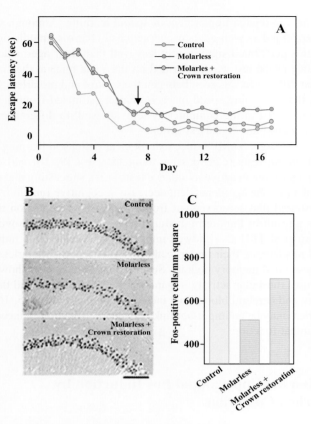

Fig. 2. Effect of placing an artificial crown on spatial learning and the number of Fos-positive cells in the CA1 region in aged molarless mice. **A** On postoperative day 10, the control group and the two molarless groups were subjected to the water maze learning test; one group of molarless mice then received an artificial crown between the tests on test days 7 and 8 (*arrow*). The results are expressed as the mean score. **B** Fos immunohistochemistry (IHC) in the CA1. IHC analysis was carried out following the learning task on the final test day (day 17). *Bar* 100 μm. **C** Quantitative results for Fos-positive cells in the CA1. Each *column* indicates the mean. (From Watanabe et al. [6], with permission)

the expression of c-fos and Fos is a useful marker for elevated levels of neuronal activity generated in the brain following various stimuli [18, 19] and the finding that trigeminal input has a facilitatory effect on synaptic transmission in various regions of the cerebral cortex [20, 50], it is assumed that deficits in spatial learning in aged molarless mice are largely due to reduced afferent input from masticatory work and mastication-associated face and jaw movements to the cerebral cortex. It has been shown that chewing increases neuronal activity and blood flow rate in various cortical regions, including the primary sensory cortex [21], and that the suppressed behavioral and hippocampal responses in appetitive trace classic conditioning are paralleled by a reduced frequency of rhythmic jaw movement [22].

Astroglial Responsiveness Under the Molarless Condition

In the hippocampus, aging has been shown to cause an increase in astrocyte numbers [23, 24]; and glial fibrillary acidic protein (GFAP), produced as an intermediate filament protein [25], has been widely used to monitor astrocyte changes in response to neuronal degeneration and aging [26] and glial reactivity during aging [27]. Thus, it seems likely that reduced mastication may be involved in astroglial responsiveness in the hippocampus.

To evaluate the mechanism(s) responsible for senile impairment of cognitive function as a result of reduced mastication, we immunohistochemically examined the effects of the molarless condition on the hippocampal expression of GFAP and on spatial memory in young adult and aged SAMP8 mice [8]. Figure 3 shows the immunohistochemical results, which indicate that the molarless condition enhanced the age-dependent increase in the density and hypertrophy of GFAP-labeled

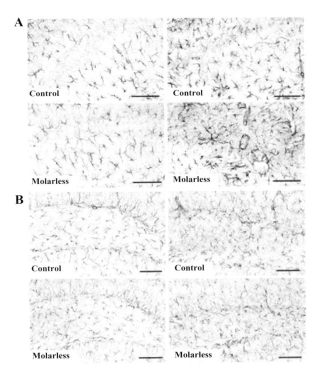

Fig. 3. Glial fibrillary acidic protein (GFAP) IHC in the CA1 region (**A**) and dentate gyrus (**B**) of control and molarless young adult (*left*) and aged (*right*) mice. Ten days after surgery, the mice began 7 days of testing in the water maze after which IHC was performed. *Bars* 150 μm. (From Onozuka et al. [8], with permission)

astrocytes in the CA1 region of the hippocampus. These effects increased as the molarless condition persisted [8]. When the extracellular K^+ concentration ($[K^+]_o$) was increased from 4 mM to 40 mM for hippocampal slices in vitro, the mean increase in the membrane potential was about 57 mV for fine, delicate astrocytes, the most frequently observed type of GFAP-positive cell in the young adult mice, and about 44 mV for the hypertrophic astrocytes of aged mice. However, there was no significant difference in resting membrane potential among these cell types. These data suggest that changes in astroglial responsiveness occur under the molarless condition in aged SAMP8 mice.

It has been suggested that the increased GFAP expression seen following deafferentation may reflect, in part, astrocyte responsiveness to changes in neuronal electrical activity [28]. For instance, Canady and Rubel demonstrated proliferation of GFAP-immunoreactive astrocytic processes in the chick cochlear nucleus following action potential blockade in the afferent nerve, indicating that neuronal activity may regulate the structure of astrocytic processes [29]. Furthermore, Rubel and MacDonald [30] have shown that an increase in GFAP-immunopositive and silver-impregnated glial processes in the chick nucleus magnocellularis occurs following cochlea removal; and they speculated that modulation of glial processes as a function of afferent activity may influence synaptic efficacy. Thus, the increase in GFAP-immunoreactive astrocytes in the CA1 region of molarless aged mice suggests a link between reduced sensory input and astroglial changes in the hippocampus.

Neuronal Degeneration in the Hippocampus Induced by the Molarless Condition

When dendritic spines in the hippocampus CA1 region of the SAMP8 mice were assessed using Golgi-Cox staining, pyramidal cells with apical and basal dendrites were seen in all mice, but the spine number was significantly decreased in aged molarless mice compared age-matched control mice (Fig. 4) [17]; this finding suggests the involvement of the molarless condition in an attenuation of input activities in hippocampal synapses. If this hypothesis is true, age-dependent neuronal cell death must be enhanced in the molarless condition. As expected, the age-dependent decrease in CA1 pyramidal neurons was enhanced by the molarless condition (Fig. 5) [8]. Quantitative analysis revealed that the percents of these neurons in the CA1 subfield of young, middle-aged, and aged molarless mice were 96.7%, 85.1%, and 77.8%, respectively, of that in age-matched control mice, indicating that a reduction in neuron numbers parallels the reduction in learning ability in a water maze test [5]. This implies that the reduction in hippocampal neuron numbers may be related to the impaired spatial memory seen in middle-aged and aged molarless mice. However, no significant difference in neuronal number was seen in either the CA3 subfield or the dentate gyrus between any of the groups.

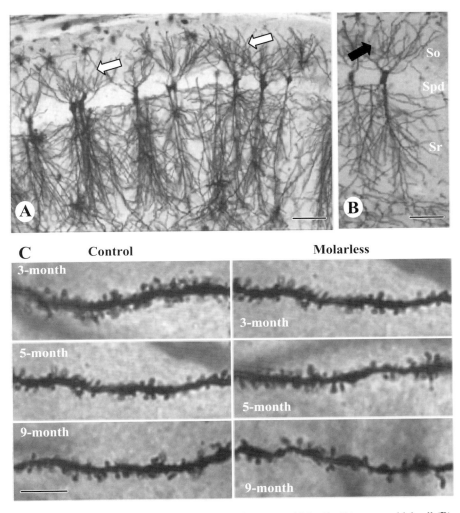

Fig. 4. Photomicrographs showing hippocampal CA1 pyramidal cells (**A**), a pyramidal cell (**B**), and dendritic spines in CA1 basal dendrites of control and molarless mice (**C**). *So*, stratum oriens; *Spd*, stratum pyramidale; *Sr*, stratum radiatum; *arrows*, basal dendrites. Bars 100 μm (**A**, **B**) and 10 μm (**C**). (From Kubo et al. [17], with permission)

Decrease in Septohippocampal Cholinergic Activities by the Molarless Condition

The cholinergic neuronal system in the hippocampus plays an important role in spatial cognition and undergoes a variety of age-dependent changes (reviewed in [31]). In rodents, hippocampal acetylcholine (ACh) release declines with age [17], and there is an age-dependent decline in memory function [32]. Furthermore,

Fig. 5. Nissl staining of the hippocampal formation of a control (*left*) or molarless (*right*) SAMP8 mouse. *DG*, dentate gyrus. *Bars* 200 μm. (From Onozuka et al. [36], with permission)

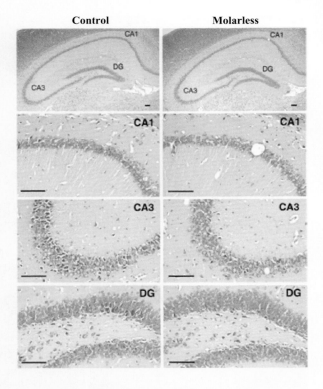

the ACh-synthesizing enzyme choline acetyltransferase (ChAT) levels in the hippocampus decrease with age [33], and cholinergic neurons in the medial septal nucleus, projecting to the hippocampal formation [34], degenerate during aging [35].

In the experiments using microdialysis, biochemical, and immunohistochemical approaches in young adult and aged SAMP8 mice, we found that the molarless condition enhances a normal age-related decrease in the functioning of the septo-hippocampal cholinergic system, which may be linked to impaired learning ability measured in a water maze [8, 36]. In addition, KCl-evoked ACh release in the hippocampus of the aged molarless group was significantly less than that in age-matched molar-intact controls (Fig. 6) [7], implying that the molarless condition suppresses hippocampal ACh release in aged SAMP8 mice. In agreement with our results, it has been reported that hippocampal ACh release in response to depolarizing stimulation using high potassium concentrations is decreased in aged rats [37], suggesting that the molarless condition in aged SAMP8 mice may be involved in the development of age-related functional impairment of the cholinergic system in the hippocampus.

In our study, aged SAMP8 mice lacking molar teeth showed a significant reduction in the number of ChAT-positive neurons in the septal nucleus compared to age-matched molar-intact SAMP8 mice (Fig. 7) [7]. In contrast, the molarless

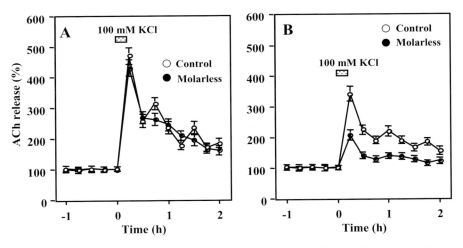

Fig. 6. Effect of the molarless condition on hippocampal acetylcholine (*ACh*) release. **A** Young adult. **B** Aged mice. Values are the mean (percent of the basal level) ± SE ($n = 5$ for each group). Basal levels were defined as the average value for the four samples taken before KCl perfusion. (From Onozuka et al. [7], with permission)

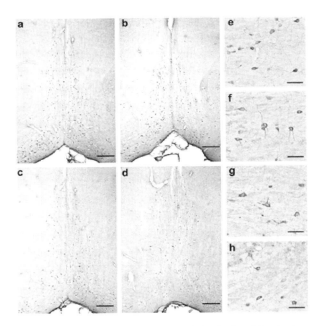

Fig. 7. Choline acetyltransferase (ChAT)-positive cells in the basal forebrain. **a, e** Young adult control. **b, f** Young adult molarless mouse. **c, g** Aged control. **d, h** Aged molarless mouse. *Bars* 200 μm (**a–d**) and 50 μm (**e–h**). (From Onozuka et al. [7], with permission)

condition had no effect on the number of ChAT-positive neurons in the vertical limb of the diagonal band of Broca (vDBB). During aging, various degenerative changes in basal forebrain cholinergic neurons have been shown in experimental animals [38, 39]. Lee et al. [40] reported a significant decrease in the number of septohippocampal cholinergic neurons in aged animals. The hippocampus is supplied by cholinergic fibers arising from the medial septal nucleus and ending on pyramidal neurons of the hippocampus and granule neurons of the dentate gyrus [41]. Taken together with the fact that an age-related impairment of cholinergic neurons in the medial septal nucleus is associated with cognitive impairment [41], it is likely that, in molarless aged SAMP8 mice, the reduced number of cholinergic neurons in the medial septal nucleus is related to impaired spatial memory [7].

Glucocorticoid Response to the Molarless Condition

Previous studies have shown that basal plasma corticosterone levels in aged rats correlate significantly with hippocampal degeneration and spatial learning deficits [23, 42]. Elevated plasma corticosterone levels are found only in aged rats with spatial memory deficits and not in those with normal spatial memory [43]. Cumulative exposure to high glucocorticoid levels throughout life disrupts electrophysiological function, leading to atrophy and ultimately the death of hippocampal neurons, all of which can cause severe cognitive deficits in hippocampus-dependent learning and memory [44]. Combined with the fact that hippocampal neuronal loss occurs in both patients with symptoms of senile dementia [45] and animals with senile deficits of cognitive function [44, 45], it is conceivable that the molarless condition-induced deficits in spatial learning and hippocampal neurons seen in aged SAMP8 may be due to the damaging effects of glucocorticoid.

As shown in Fig. 8, the corticosterone levels showed significant circadian variation, peaking in both groups at the onset of the dark period (i.e., 8 p.m.). However,

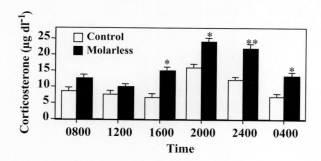

Fig. 8. Effects of the molarless condition on plasma corticosterone levels in aged SAMP8 mice. Mean ± SE plasma corticosterone levels in control and molarless mice ($n = 4$ for each column) at various times over a 24-h cycle. *$P < 0.05$ compared with controls; **$P < 0.01$ compared with controls. (From Onozuka et al. [47], with permission)

at all time points, the molarless group had significantly higher plasma corticosterone levels, indicating that the molarless condition causes increased plasma corticosterone levels. This finding was similar to that reported in a previous study [46] in showing a peak at the beginning of the dark period, when activity is generally greatest, and the lowest levels near the end of the dark period or the beginning of the light period, when rodents are least active. However, the molarless group had overall higher corticosterone levels than the control group, indicating that the molarless condition results in increased exposure to corticosterone and suggesting impaired hypothalamic-pituitary-adrenal negative feedback inhibition in molarless mice.

We next wanted to assess the effect of the corticosterone synthesis inhibitor metyrapone on the molarless condition-induced increase in corticosterone levels and reduction of CA1 neurons. Therefore, 1 day before the operation and every 2 days until the collection of blood, molarless mice received an injection of this inhibitor at a dosage known to inhibit the stress-induced rise in plasma corticosterone levels and hippocampal neuronal damage [48]. Ten days after the operation, we measured plasma corticosterone levels in the control ($n = 10$), molarless, and metyrapone-treated molarless groups. As shown in Fig. 9, it suppressed the molarless condition-induced increase in plasma corticosterone levels (a, in Fig. 9) but had no significant effect on plasma corticosterone levels in control mice [49]. Also, metyrapone prevented the molarless-induced reduction in neuronal number in this subfield (b and c, in Fig. 9), as no significant difference in the number was seen between vehicle-injected control mice and metyrapone-injected molarless mice. Furthermore, metyrapone prevented the increase in escape latency in the water maze test induced by the molarless condition, as a difference in latency was seen between metyrapone-injected molarless mice and vehicle-injected molarless mice but not between metyrapone-injected molarless mice and vehicle-injected control mice (d and e, in Fig. 9), implying that the molarless-induced deficits in spatial learning and hippocampal neuron numbers in aged SAMP8 mice may be related to exposure to increased corticosterone levels. Together with the observations that the adrenals of molarless aged SAMP8 mice are heavier than those of age-matched molar-intact control mice [8] and that increased mastication reduces the plasma corticosterone response during novelty exposure in mice [49], a prolonged stressful response to the molarless condition and the consequent increase in exposure to corticosterone could hasten hippocampal neuron damage in this species.

Conclusion

Based on the above findings, we propose that the reduced ability to masticate in aged SAMP8 mice induces deficits in spatial cognitive memory. The mechanism(s) underlying this phenomenon are as follows: (1) a decrease in input activities in the hippocampus; (2) degeneration of hippocampal neurons; (3) a decrease in the

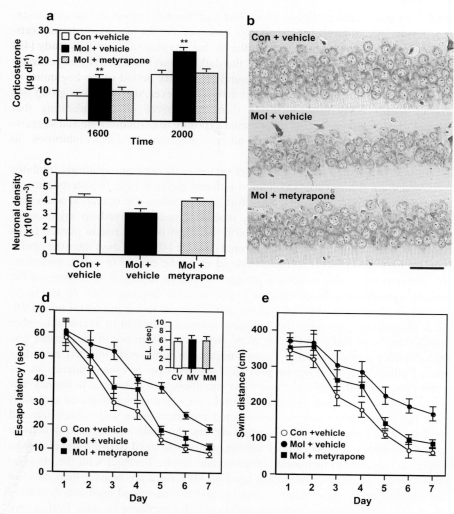

Fig. 9. Effects of metyrapone. **a** Mean ± SE ($n = 10$ for each column) plasma corticosterone levels in vehicle-injected control, vehicle-injected molarless, and metyrapone-injected molarless groups at 1600 and 2000h (4 p.m. and 8 p.m.) **$P < 0.01$ compared with vehicle-injected controls. **b** Representative photomicrographs of cresyl violet-stained sections showing the CA1 cell field in the control (*top*), molarless (*middle*) and metyrapone-treated molarless (*bottom*) groups. *Bar* 50μm. **c** Mean ± SE ($n = 7$ for each column) neuron density in the CA1 pyramidal cell field in vehicle-injected control, vehicle-injected molarless, and metyrapone-injected molarless groups ($n = 7$ animals per group). *$P < 0.05$. **d, e** Spatial learning in a water maze test. Ten days after surgery, the learning test was started. Mean ± SE ($n = 10$ for each group) latency (**d**) or swim distance (**e**) to locate a submerged platform in a Morris swim maze in vehicle-injected control, vehicle-injected molarless, and metyrapone-injected molarless groups. **Inset** Visible probe test. At the end of the maze test, animals performed a visible probe test. *Con*, molar-intact control group; *Mol*, molarless group; *E.L.*, escape latency; *CV*, vehicle-injected control group; *MV*, vehicle-injected molarless group; *MM*, metyrapone-injected molarless group. There was no significant difference between groups. (From Onozuka et al. [47], with permission)

septohippocampal cholinergic networks; and (4) involvement of the glucocorticoid response in the hippocampus. Therefore, we strongly suggest that dysfunctional mastication is one of the risk factors for senile dementia.

Acknowledgments Our work introduced here was supported by Grants-in-Aid for Scientific Research from the Ministry of Education, Science, and Culture of Japan and a grant from Lotte Co. Ltd. (Tokyo, Japan).

References

1. Kondo K, Niino M, Shido K (1995) A case-control study of Alzheimer's disease in Japan: significance of life-styles. Dementia 5:314–326.
2. Kawamura S (1989) The effect of food consistency on conditioned avoidance response in mice and rats. Jpn J Oral Biol 31:71–82.
3. Fujisawa H (1990) The effect of mastication on post-natal development of the rat brain with a histological and behavioral study. J Jpn Oral Biol 32:495–508.
4. Morris RGM (1984) Development of a water-maze procedure for studying spatial learning in the rat. J Neurosci Methods 11:47–60.
5. Watanabe K, Tonosaki K, Kawase T, et al. (2001) Evidence for involvement of dysfunctional teeth in the senile process in the hippocampus of SAMP8 mice. Exp Gerontol 36:283–295.
6. Watanabe K, Ozono S, Nishiyama K, et al. (2002) The molarless condition in aged SAMP8 mice attenuates hippocampal Fos induction linked to water maze performance. Behav Brain Res 128:19–25.
7. Onozuka M, Watanabe K, Fujita M, et al. (2002) Changes in the septohippocampal cholinergic system following removal of molar teeth in the aged SAMP8 mouse. Behav Brain Res 133:197–204.
8. Onozuka M, Watanabe K, Nagasaki S, et al. (2000) Impairment of spatial memory and changes in astroglial responsiveness following loss of molar teeth in aged SAMP8 mice. Behav Brain Res 108:145–155.
9. Miyamoto M, Kiyota Y, Yamazaki N, et al. (1986) Age-related changes in learning and memory in the senescence-accelerated mouse (SAM). Physiol Behav 38:399–406.
10. Yagi H, Katoh S, Akiguchi I, et al. (1988) Age-related deterioration of ability of acquisition in memory and learning in senescence accelerated mouse: SAM-P/8 as an animal model of disturbance in recent memory. Brain Res 474:86–93.
11. Flood JF, Morley JE (1993) Age-related changes in foot shock avoidance acquisition and retention in senescence accelerated mouse (SAM). Neurobiol Aging 14:153–157.
12. Ohta A, Hirano T, Yagi H (1989) Behavioral characteristics of the SAM-P/8 strain in Sidman active avoidance task. Brain Res 498:195–198.
13. Kubota K, Nagae K, Shibanai S (1988) Degenerative changes of primary neurons following tooth extraction. Anat Anz Jena 166:133–139.
14. Kato T, Usami T, Noda Y (1997) The effect of the loss of molar teeth on spatial memory and acetylcholine release from the parietal cortex in aged rats. Behav Brain Res 83:239–242.
15. Abram R, Hammel HT (1964) Hypothalamic temperature in unanesthetized albino rats during feeding and sleeping. Am J Physiol 206:641–646.
16. Rampone AJ, Shirasu ME (1964) Temperature changes in the rat in response to feeding. Science 144:317–331.
17. Kubo K, Iwaku F, Watanabe K, et al. (2005) Molarless-induced changes of spines in hippocampal region of SAMP8 mice. Brain Res 1057:191–195.

18. Bialy M, Kaczmarek L (1996) c-Fos expression as a tool to search for the neurobiological base of the sexual behaviour of males. Acta Neurobiol Exp 56:567–577.
19. Dragunow M, Faull R (1989) The use of c-fos as a metabolic marker in neuronal pathway tracing. J Neurosci Methods 29:261–265.
20. Morilak DA, Jacobs BL (1985) Noradrenergic modulation of sensorimotor processes in intact rats: the masseteric reflex as a model system. J Neurosci 5:1300–1306.
21. Momose I, Nishikawa J, Watanabe T, et al. (1997) Effect of mastication on regional cerebral blood flow in humans examined by positron-emission tomography with ^{15}O-labelled water and magnetic resonance imaging. Arch Oral Biol 42:57–61.
22. Seager MA, Asaka Y, Berry SD (1999) Scopolamine disruption of behavioral and hippocampal responses in appetitive trace classical conditioning. Behav Brain Res 100:143–151.
23. Landfield PW, Lindsey JD, Lynch G (1978) Hippocampal aging and adrenocorticoids: quantitative correlations. Science 202:1098–1102.
24. Lindsey JD, Landfield PW, Lynch G (1979) Early onset and topographical distribution of hypertrophied astrocytes in hippocampus of aging rats: a quantitative study. J Gerontol 34:661–671.
25. Steinert PM, Roop DR (1989) Molecular and cellular biology of intermediate filaments. Annu Rev Biochem 57:593–625.
26. Eng LF, Chirnikar RS (1994) GFAP and astrogliosis. Brain Pathol 4:229–238.
27. Landfield PW, Rose G, Sandles L, et al. (1977) Pattern of astroglial hypertrophy and neuronal degeneration in the hippocampus of aged, memory-deficient rats. J Gerontol 32:3–12.
28. Canady KS, Hyson RL, Rubel EW (1994) The astrocytic response to afferent activity blockade in chick nucleus magnocellularis is independent of synaptic activation, age and neuronal survival. J Neurosci 14:5973–5985.
29. Canady KS, Rubel EW (1992) Rapid and reversible astrocytic reaction to afferent activity blockade in chick cochear nucleus. J Neurosci 12:1001–1009.
30. Rubel EW, MacDonald GH (1992) Rapid growth of astrocytic processes in n. magnocellularis following cochlea removal. J Comp Neurol 318:415–425.
31. Berger-Sweeney J, Stearns NA, Murg SL, et al. (2001) Selective immunolesions of cholinergic neurons in mice: effects on neuroanatomy, neurochemistry, and behavor. J Neurosci 21:8164–8173.
32. Hellweg R, Fisher W, Hock C, et al. (1990) Nerve growth factor levels and choline acetyltransferase activity in the brain of aged rats with spatial memory impairments. Brain Res 537:123–130.
33. Springer JE, Tayrien MW, Loy R (1987) Regional analysis of age-related changes in the cholinergic system of the hippocampal formation and basal forebrain of the rat. Brain Res 407:180–184.
34. Amaral DG, Kurz J (1985) An analysis of the origins of the cholinergic and noncholinergic projections to the hippocampal formation of the rat. J Comp Neurol 240:37–59.
35. Wenk GL, Willard LB (1998) The neural mechanisms underlying cholinergic cell death within the basal forebrain. Int J Dev Neurosci 16:729–735.
36. Onozuka M, Watanabe K, Mirbod SM, et al. (1999) Reduced mastication stimulates impairment of spatial memory and degeneration of hippocampal neurons in aged SAMP8. Brain Res 826:148–153.
37. Takei N, Nihonmatsu I, Kawamura H (1989) Age-related decline of acetylcholine release evoked by depolarizing stimulation. Neurosci Lett 101:182–186.
38. Isacson O, Sofroniew MV (1992) Neuronal loss or replacement in the injured adult cerebral neocortex induces extensive remodeling of intrinsic and afferent neural systems. Exp Neurol 117:151–175.
39. Minger SL, Davies P (1992) Persistent innervation of the rat neocortex by basal forebrain cholinergic neurons despite the massive reduction of cortical target neurons. I. Morphometric analysis. Exp Neurol 117:124–138.
40. Lee JM, Ross ER, Gower A, et al. (1994) Spatial learning deficits in the aged rat: neuroanatomical and neurochemical correlates. Brain Res Bull 33:489–500.

41. Fischer W, Wictorin K, Björklund A (1989) Degenerative changes in forebrain cholinergic nuclei correlate with cognitive impairments in aged rats. Eur J Neurosci 2:34–45.
42. Landfield PW, Baskin RK, Pitler TA (1981) Brain aging correlates: retardation by hormonal-pharmacological treatments. Science 214:581–583.
43. Issa AM, Rowe W, Gauthier S, et al. (1990) Hypothalamic-pituitary-adrenal activity in aged, cognitively impaired and cognitively unimpaired rats. J Neurosci 10:3247–3254.
44. Squire LR (1992) Memory and the hippocampus: a synthesis from findings with rats, monkeys, and humans. Psychol Rev 99:195–231.
45. Sirevaag AM, Black JE, Greenough WT (1991) Astrocyte hypertrophy in the dentate gyrus of young male rats reflects variation of individual stress rather than group environmental complexity manipulations. Exp Neurol 111:74–79.
46. Manning JM, Bronson FH (1991) Suppression of puberty in rats by exercise: effects on hormone levels and reversal with GnRH infusion. Am J Physiol 260:R717–R723.
47. Onozuka M, Watanabe K, Fujita M, et al. (2002) Evidence for involvement of glucocorticoid response in the hippocampal changes in aged molarless SAMP8 mice. Behav Brain Res 131:125–129.
48. Krugers HJ, Maslam S, Korf J, et al. (1998) The corticosterone synthesis inhibitor metyrapone prevents hypoxia/ischemia-induced loss of synaptic function in the rat hippocampus. Stroke 2:237–249.
49. Hennessy MB, Foy T (1987) Nonedible material elicits chewing and reduces the plasma corticosterone response during novelty exposure in mice. Behav Neurosci 101:237–245.
50. Ellrich J, Andersen OK, Messlinger K, et al. (1999) Convergence of meningeal and facial afferents onto trigeminal brainstem neurons: an electrophysiological study in rat and man. Pain 82:229–237.

Cellular and Molecular Aspects of Short-Term and Long-Term Memory from Molluscan Systems

Manabu Sakakibara

Summary

Cellular and molecular mechanisms of short-term memory and long-term memory are reviewed based on observations of molluscan models of *Aplysia californica*, *Lymnaea stagnalis*, and *Hermissenda crassicornis*. It is generally accepted that short-term memory results from changes in the synaptic strength of preexisting neuronal connections that involve covalent modifications of preexisting proteins by various kinases. On the other hand, the synaptic plasticity underlying long-term memory is believed to involve protein synthesis and modulation of gene expression to induce new mRNA, protein synthesis, and morphologic modifications. These processes and mechanisms are compared in three molluscan model systems and likely have commonalities with those of mammals.

Key words Short-term memory, Long-term memory, Synaptic strength, Protein synthesis, Gene expression

Introduction

Although recent progress in molecular biology enables us to manipulate genes in mammals for a better understanding of higher brain function, molluscan models are still useful for studying the underlying mechanisms of brain function. It is difficult to study how synaptic plasticity produces a change in behavior in mammalian preparations owing to the large number of neurons and synapses involved in producing the behavior. Gastropod molluscs are established animal models for studying the neuronal mechanisms of learning and memory because the synaptic plasticity underlying changes in their behavior are easily observed.

Laboratory of Neurobiological Engineering, Department of Biological Science and Technology, School of High Technology for Human Welfare, Tokai University, 317 Nishino, Numazu 410-0321, Japan

This chapter is dedicated to the late Professor Herman T. Epstein.

The mammalian brain has two memory systems: One is declarative, and the other is nondeclarative. Declarative memory is sometimes referred to as "explicit memory," or the conscious recall of knowledge, which is well developed in mammals and is dependent on cerebral cortical structures, including the hippocampus. In contrast, nondeclarative memory, sometimes referred to as "implicit memory," is memory for motor skills and involves the cerebellum and striatum in the mammalian brain. The learning mechanism studied in invertebrate models corresponds to that of nondeclarative memory.

In the mammalian brain, cellular and molecular changes that occur during the formation of both types of memory, declarative and nondeclarative, are difficult to study because the contribution of various synapses is not clearly identified. The invertebrate model system is useful for bridging this gap because the cellular and molecular analyses of behavioral problems using this simpler system facilitates our understanding of the synaptic loci and underlying fundamental mechanisms of learning and memory in general. The model animals exhibit several forms of learning—habituation, sensitization, classical conditioning, operant conditioning—which include many of the behavioral features of learning in mammals, suggesting that learning in molluscs and mammals has common mechanisms.

The modern physiology of learning and memory began during the early twentieth century with Pavlov's pioneering studies [1]. Canadian psychologist Donald Hebb published a theory of brain function and learning during the mid-twentieth century; it was a modified switchboard theory. A given memory was represented by a set of neurons that had developed increased functional connections. The basic idea was that new learning is fragile, and well-established memories are not; this process of establishing long-lasting memories is termed *consolidation*. Hebb proposed that if a neural connection interacted in a certain manner lasting cellular change would occur and an association would be formed, thereby providing the substrate of memory.

> When an axon of cell A is near enough to excite a cell B and repeatedly or persistently takes part in firing it, some growth process or metabolic change takes place in one or both cells such that A's efficiency, as one of the cells firing B, is increased.—Hebb [2]

Short-Term Memory, Long-Term Memory

This chapter focuses mainly on the cellular and molecular mechanisms of short-term memory (STM) and long-term memory (LTM) based on recent studies of gastropod molluscs, such as *Aplysia, Lymnaea*, and *Hermissenda*.

It is usually thought that STM lasts for minutes, whereas LTM lasts for days, weeks, years, or even as long as the entire life-span. There is also an intermediate-term memory (ITM) that falls between STM and LTM and lasts for ≥ 1 h. The cellular and molecular aspects of STM suggest that STM is due to a change in the synaptic strength of preexisting neuronal connections through covalent modification of preexisting proteins by various kinases such as protein kinase A (PKA), protein kinase C (PKC), calcium/calmodulin-dependent protein kinase II (CaMK II), and mitogen-activated protein kinase (MAPK). LTM, in contrast, is thought to involve the modulation of gene expression to induce new mRNA and protein synthesis. It is hypothesized that ITM requires new protein synthesis from preexisting transcriptional factors but does not require new protein synthesis from new mRNA synthesis, as is the case for LTM. ITM is sometimes referred to as the transition state from STM to LTM because although genetic translation is required genetic transcription is not [3, 4]. At the ITM to LTM stage, new structural alterations at the synapse are observed, such as the growth of new synaptic connections or synaptic remodeling. To analyze dynamic changes at the identified synapse, a gastropod model system is desirable and must be successfully conditioned with one-trial conditioning for in vivo study or must be observable in a network made up of dissociated cell cultures for in vitro study. The learned behavior must be such that there is a distinct time window of how long the memory persists. Our research strategy is to identify the time window of the transition from STM to LTM and then examine the effects of blocking genetic translation and transcription from the viewpoint of biophysics, morphology, and molecular biology.

The learning ability of each individual animal is quite different; some animals learn quickly and have good long-lasting memory even if they experience only one conditioning trial, whereas others have poor learning and memory performance even after several conditioning trials. These interindividual differences are thought to be due to individual differences in motivation that are dependent on the activity of the amygdala in mammals; however, these motivational differences exist even in invertebrates that lack a brain. In invertebrates, the physical condition, such as the state of starvation, influences their motivation, especially in feeding behavior-related conditioning paradigms. The differences between good and poor performers in terms of the conditioning paradigm and motivation are discussed below.

Aplysia californica

Aplysia is a well-established animal model for studying the synaptic plasticity underlying the gill- and siphon-withdrawal reflex. With this reflex, the animal shows several forms of learning, such as habituation, sensitization, and classical conditioning [5, 6]. The synapse between the sensory neuron and the motor neuron is the site of plasticity. Behavior in response to molecular mechanisms has been studied in isolated, semi-intact, and dissociated cell culture preparations. The molecular mechanisms contributing to implicit memory storage have been most extensively studied with the *Aplysia* gill- and siphon-withdrawal reflex [7].

Aplysia has a simple nervous system containing approximately 20 000 neurons and displays a variety of defensive reflexes for withdrawing its tail, gill, and siphon. A light touch to the siphon elicits withdrawal of both the siphon and gill, whereas a tactile stimulus to the tail elicits only tail withdrawal. These reflex withdrawals habituate with repeated stimulation. In response to a newly encountered stimulus to the siphon, the sensory neuron innervating the siphon generates excitatory synaptic potentials in the interneurons and motor neurons. These synaptic potentials integrate spatiotemporally and strongly excite the motor neurons, leading to strong withdrawal of the gill. If the stimulus is presented repeatedly, the synaptic potentials produced by the sensory neurons in the interneurons and motoneurons become progressively smaller. The synaptic potentials in the motor neurons produced by some of the excitatory interneurons also become weaker, which results in a reduction in the strength of the reflex response. This reduction in the reflex response is termed *habituation* [8]. The decreased synaptic transmission in the sensory neurons results from a decrease in the amount of a chemical neurotransmitter, in this case glutamate, released from the synaptic terminals. When an animal repeatedly encounters a harmless stimulus it learns to habituate to the stimulus, whereas when the animal is exposed to a harmful stimulus it learns to respond more vigorously not only to the stimulus but also to other harmless stimuli. This reflex enhancement is termed *sensitization*. Sensitization is an elementary form of nonassociative learning by which *Aplysia* acquires information about the properties of a single noxious stimulus. As with other forms of defensive behavior, the memory for sensitization of the withdrawal reflex is graded, and repeated tail shocks lead to a long-lasting memory. A single tail shock produces short-term sensitization that lasts for minutes, and repeated tail shock produces long-term sensitization that lasts for up to 1 week [9]. Figure 1a shows a simplified diagram of habituation, and Fig. 1b shows a circuit diagram of short-term sensitization in *Aplysia*.

Sensory neurons in the abdominal ganglion innervating the siphon skin use glutamate as the neurotransmitter and terminate on motor neurons that innervate the gill. Stimuli to the tail activate sensory neurons that excite facilitating interneurons. Serotonin (5-hydroxytryptamine, or 5-HT) is the neurotransmitter of the facilitating interneuron, which forms a synapse on the terminal of the sensory neuron innervating the siphon skin and motor neuron controlling gill withdrawal (Fig. 1A). Habituation leads to homosynaptic depression, a decrease in synaptic strength resulting from sustained direct activity in the sensory neuron. On the other hand, sensitization involves heterosynaptic facilitation; that is, the sensitizing stimulus activates a group of interneurons that form synapses on the sensory neurons. Because the mechanism of sensitization in *Aplysia* has been well studied and is described in detail elsewhere (e.g., in a textbook [10] and in reviews [7, 11–13]), only a brief summary of the in vivo and in vitro systems is provided here.

In a dissociated culture network of sensory and motor neurons, a brief application of 5-HT, a modulatory transmitter normally released from the facilitating interneuron by sensitizing stimuli in the intact animal, mimics a tail shock, leading to short-term facilitation of the motor neurons [14]. If 5-HT is applied intermittently at some interval, however, it induces long-term facilitation [15]. Furthermore, if the sensory neurons as presynaptic elements are in an excitatory state before the 5-HT is applied, the facilitating effect induced by repeated 5-HT application is larger and longer lasting. This phenomenon is the same as classical conditioning [16]. Serotonin binds to cell surface receptors on the sensory neurons and facilitates the production of a diffusible second messenger, cyclic adenosine monophosphate (cAMP), by activating adenylyl cyclase. The increase in cytosolic cAMP results in

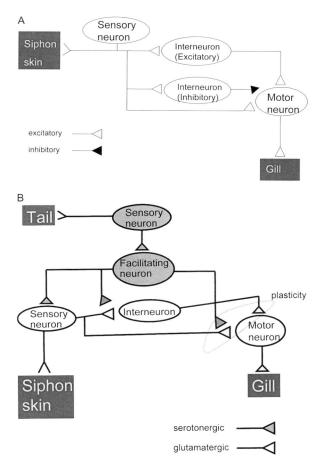

Fig. 1. A Simplified circuit involved in the gill-withdrawal reflex. In this circuit, mechanosensory neurons innervate the siphon skin. These sensory cells use glutamate as their neurotransmitter and terminate on a cluster of six motor neurons that innervate the gill and on several groups of excitatory and inhibitory interneurons that synapse on the motor neurons. Repeated stimulation of the siphon leads to depression of synaptic transmission between the sensory and motor neurons as well as between certain interneurons and motor neurons. **B** Sensitization of the gill is produced by applying a noxious stimulus to another part of the body. Stimuli to the tail activate sensory neurons that excite facilitating interneurons. The facilitation cells, some of which use serotonin (5-hydroxytryptamine, or 5-HT) as their neurotransmitter, form synapses on the terminal of the sensory neurons innervating the siphon skin. There they enhance transmitter release from the sensory neurons by means of presynaptic facilitation

short-term facilitation. This facilitation is due to enhancing the release of the neurotransmitter from the sensory receptor terminal, thus inactivating K^+ channels [9], and enhanced excitability. All of the mechanisms mentioned above are presynaptic events; that is, they occur in the sensory neuron. At the sensory-to-motor neuron synapse, the degree of sensitization seems to depend on both the duration of 5-HT exposure and the state of the synapse—that is, whether the synapse is in a resting or a depressed state. Many studies have demonstrated that brief exposure to 5-HT activates PKA in the sensory neuron, leading to inactivation of K^+ channels, an increase in intracellular Ca^{2+}, and transmitter release [17–19]. Longer exposure to 5-HT activates PKC and CaMK II in the sensory neuron [20, 21]. In addition, longer exposure to 5-HT leads to intermediate-term facilitation, which requires protein synthesis [20]. Exposure to 5-HT for more than 5 min might involve postsynaptic mechanisms such as inositol trisphosphate-mediated CaMK II activation and α-amino-3-hydroxy-5-methyl-4-isoxazolepropionic acid (AMPA) receptor insertion [22, 23], as well as presynaptic mechanisms such as simultaneous PKA and PKC activation. These findings suggest that not only the specific kinase involved but its site of action might depend on the duration of transmitter exposure, in this case 5-HT; brief exposure affects the presynaptic mechanism, whereas long exposure involves both presynaptic and postsynaptic mechanisms.

Activity-dependent plasticity in *Aplysia* might involve both pre- and postsynaptic mechanisms. At the synapse between sensory and motor neurons, presynaptic tetanic stimulation evokes long-term potentiation similar to that observed in the mammalian hippocampus. This potentiation involves metabotropic glutamate receptors, and potentiation decreases after injection of Ca^{2+} into the sensory neuron or injection of a CaMK II inhibitor into the motor neuron [23, 24]. These findings indicate that potentiation involves both pre- and postsynaptic mechanisms that interact with each other (i.e., with a strong shock to the tail both postsynaptic Ca^{2+} and CaMK II and presynaptic PKA contribute to sensitization, whereas sensitization with a weak shock is an entirely presynaptic event). Classical conditioning experiments in *Aplysia* with a siphon touch [conditioned stimulus (CS)] and tail shock [unconditioned stimulus (US)] revealed that both pre- and postsynaptic mechanisms are involved [25].

Briefly, with STM, synaptic stimulation of sensory neurons leads to a local increase in cAMP and activation of PKA by causing the catalytic subunits of the enzyme to dissociate from the regulatory subunits. The catalytic subunits then phosphorylate K^+ channels and proteins to enhance transmitter release. In contrast, repeated synaptic stimulation resulting in a persistent increase of cAMP levels leads to long-term synaptic plasticity.

In addition to activating various protein kinases after STM, protein phosphatases such as calcineurin and protein phosphatase 1, acting as inhibitory constraints of memory formation, are also suggested to have a key role in regulating LTM [26]. It is generally assumed that an equilibrium state between kinase and phosphatase activities at a given synapse is critical to gate the synaptic signal reaching the nucleus to stabilize memory formation and retrieval. LTM is represented at the cellular level by activity-dependent modulation of both the function and structure of specific synaptic connections, which in turn depend on activation of a specific pattern of gene expression [7]. PKA activates gene expression by phosphorylating transcription factors that bind to the cAMP-responsive element (CRE). The CRE is one of the DNA response elements contained in the control region of the gene. The binding of various transcription factors to these response elements regulates the activity of RNA polymerase, thereby determining when and to what level a gene is expressed. One of the major factors that recognize the CRE is a CRE-binding protein (CREB1), a transcriptional activator. Dash et al. demonstrated that PKA activates gene expression through CREB during the formation of LTM. If CREB1 is essential for LTM, selective blockade of CREB1 should eliminate LTM formation. A CRE oligonucleotide injected into a sensory neuron co-cultured with motor neurons inhibits the function of CREB1 by binding to the CREB1 protein in the cell [27]. Various transcriptional enzymes of the CREB family are involved in LTM formation; CREB1 acts as an activator, and CREB2 acts as a repressor. It seems that a balance between the CREB activator and repressor is important for LTM formation. Overexpression of an inhibitory form blocks LTM but not STM, whereas overexpression of an activator has the opposite effect and increases the efficacy of training in LTM formation.

After the LTM-related structural modification occurs at the presynaptic sensory neuron varicosities due to sensitization, sensory neurons exhibit a twofold increase in the total number of synaptic varicosities and in the size of each neuron's arbor [28–30]. Bailey and Chen reported that, after behavioral extinction of sensitization, changes in the varicosities and active zone number persisted for at least 1 week and were partially reversed by the end of the 3-week experiment [30]. Kim et al. observed functional and presynaptic structural changes during long-term facilitation with time-lapse confocal microscopic imaging. Long-term facilitation results in structural changes in presynaptic neurons. These findings suggest two possible mechanisms: activation of preexisting silent presynapses through filling with synaptic vesicles or the generation of new synaptic varicosities. The activation of preexisting silent presynapses, a rapid process that occurs within 3–6 h after facilitation, requires only translation, whereas the generation of new varicosities is a comparatively slow process that occurs within 24 h and requires both translation and transcription [31].

Lymnaea stagnalis

A number of classic conditioning [32–36] and operant conditioning [37, 38] paradigms have been used successfully in *Lymnaea*, and cellular traces of behavioral conditioning have been identified in isolated, semi-intact, and simplified culture networks. In this model system, associative learning of appetitive or aversive conditioning of the feeding behavior and aversive operant conditioning of respiratory behavior are well characterized. To understand the conditioning-induced modification of the neural networks underlying feeding behaviors, four groups of neurons— sensory, modulatory, central pattern generator, and motor neurons—have been studied, and the components of the neurons in each group have been analyzed at the cellular and molecular levels.

Previous studies examined the chemosensory neuron network—cerebral giant cells (CGC), slow oscillator cells (SO), and cerebral ventral 1 cells (CVl) in the modulatory neuron group; N1, N2, and N3 in the central pattern generator neuron group; and B7, B3, and B4 in the motor neuron group—leading to sequences of muscle activity and feeding movements [39, 40] for appetitive conditioning. It is hypothesized that CGC and B2 motor neurons have key roles in mediating aversive conditioning [34, 41]. Three neuron groups are involved in the aversive operant conditioning of respiratory behavior: sensory neurons, respiratory central pattern generator, and motoneurons controlling the pneumostome muscle. Sensory neurons located in the pneumostome-osphradial area activate the respiratory central pattern generator, comprised of the right pedal dorsal 1 (RPeDl), visceral dorsal 4 (VD4), and input 3 (IP3) neurons. VD4 drives activity in the K motoneurons responsible for pneumostome closing, and IP3 drives activity in the I/J motoneurons responsible for pneumostome opening [42]. Studies of this operant conditioning preparation indicate that RPeD1 is the necessary site for LTM formation and memory storage because following soma ablation the neural circuit is capable only of mediating learning and ITM. LTM cannot be demonstrated after soma ablation. Thus, the soma, where the new mRNA is synthesized to make new protein for LTM formation, is the necessary site as it functions as a protein-synthesizing factory [4, 42, 43]. Figure 2 shows circuit diagrams and neurons involved in feeding behavior and respiratory behavior in *Lymnaea*.

In the appetitive conditioning paradigm, light mechanical touch around the lip is a CS, and sucrose is a US [39]. For aversive conditioning, sucrose is a CS, and aversive KCl application or moderate mechanical touch to the head acts as a US [35, 44]. Both one-trial appetitive and aversive conditioning have been successfully performed, resulting in the formation of LTM [45–49]. A single appetitive conditioning trial results in memory that persists at least 21 days [46], and the memory of a single aversive conditioning trial persists at least 24 h to 7 days [49]. It is interesting that the motivation of the animals differs between the appetitive and aversive conditioning paradigms [49, 50]. For appetitive conditioning, the animals were food-deprived for 4 days prior to the conditioning, and this condition was sufficient for good performance [46, 50]. In contrast, food deprivation for 5 days did not motivate animals, as indicated by good retention; rather, CS-sucrose causes cessation of feeding behavior in more than half of the animals tested [49]. It is unlikely that 5 days of food deprivation made the animals sick; instead, a long period of food deprivation might have induced excessive stress, which decreased their motivation.

It is generally assumed that there are two critical periods for LTM formation and for memory recall; one is activated soon after the conditioning trial to synthesize protein from preexisting transcriptional factors, and the other is activated later to synthesize protein from new transcriptional factors. In accordance with this hypothesis, injection of the translation blocker anisomycin 2.5 h before training prevents the formation of ITM (lasting 1–3 h) and LTM (lasting > 6 h). On the other

Feeding behavior

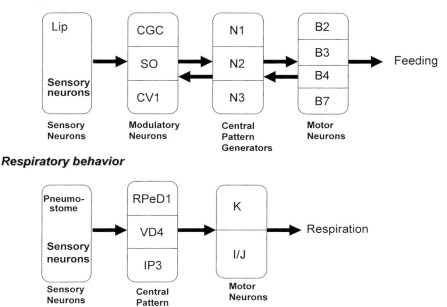

Fig. 2. Feeding and respiratory networks of *Lymnaea stagnalis*. **Top panel** Neuronal network underlying feeding behavior. Chemosensory neurons located in the lip detect the presence of food. The sensory neurons send their information to modulatory neuron groups: cerebral giant cells (*CGC*), slow oscillator cells (*SO*), and cerebral ventral 1 cells (*CV1*). The rhythmic pattern of the central pattern generators produces the feeding cycle of protraction, rasp, and swallow phase, leading to muscular activity of B2, B3, B4, and B7 motor neurons. **Bottom panel** Neuronal network underlying respiratory behavior. Activation of sensory neurons located in the pneumostome-osphradial area leads to activation of the respiratory central pattern generators comprising right pedal dorsal 1 (*RPeD1*), ventral dorsal 4 (*VD4*), and input 3 (*IP3*) neurons. VD4 drives activity in K motoneurons, and IP3 drives activity in I/J motoneurons, which are responsible for pneumostome closing and opening, respectively

hand, injection of the transcription blocker actinomycin-D 2.5 h before training did not prevent the establishment of ITM but blocked LTM formation. Thus, in *Lymnaea*, following aversive operant conditioning both ITM and LTM are dependent on new protein synthesis [4]. In addition to pharmacologic blockade using anisomycin and actinomycin D, translation and transcription factors can be physically manipulated. Immediately after "taste avoidance conditioning" in *Limax*, Sekiguchi et al. exposed animals to 1°C for 1 h to induce retrograde amnesia [51]. Using this cooling technique in the respiratory operant conditioning of *Lymnaea*, Sangha et al. reported that cooling the animals for 1 h immediately after training was sufficient to block both ITM and LTM, whereas cooling them for a similar period starting 10 or 15 min after cessation of training failed to block ITM or LTM formation, respectively [52]. Further cooling extended LTM that normally persisted for 2 days to at least 8 days [52], demonstrating that cooling prevents forgetting, and thus forgetting is an active process that is not part of the memory consolidation process. This operant conditioning can be extinguished with the spaced backward conditioning procedure more effectively than with massed conditioning trials. The memory was extinguished within 1 h after the extinction conditioning trial; the extinction was due to new mRNA and protein synthesis at the soma of the RPeDl, which is required for LTM consolidation [53]. A series of experiments by Sangha et al. demonstrated that there are two critical periods required for LTM: one immediately after conditioning and the other several hours later. This two-stage protein synthesis theory, however, was not supported in the one-trial appetitive conditioning paradigm [45].

Pharmacological blockade by injection 10 min after conditioning of either the translation inhibitor anisomycin or the transcription inhibitor actinomycin-D blocked LTM. Further anisomycin injection 1, 2, 3, 4, 5, and 6 h after the conditioning paradigm had no effect on memory recall. These results are in contrast to those from the aversive operant conditioning mentioned above and indicate that there is only a single critical period between 10 min and 1 h for protein synthesis in appetitive conditioning. These differences might reflect the involvement of a different neuronal network in each conditioning paradigm. Consolidation is believed to involve the regulation of genes involved in long-term stabilization of synaptic modifications in the neuronal circuits activated during learning.

A critical step in this process involves the activation of immediate early genes (IEGs), which are rapidly induced activity-dependent genes that encode transcription factors capable of regulating the transcription of a number of downstream late-responding genes. It is generally assumed that the crucial point for memory consolidation is the timing of IEG activation. Several IEGs—such as CCAAT/enhancer-binding protein (C/EBP), activity-related cytoskeleton associated protein (Arc), c-fos, and c-jun—have been studied with regard to the stabilization of long-lasting synaptic plasticity and LTM formation [54–58]. Recent findings indicate that LTM formation in aversive conditioning for feeding behavior involves a combination of C/EBP synthesis and phosphorylation as well as C/EBP mRNA breakdown in the pair of B2 motoneurons that control feeding behavior [59].

Hermissenda crassicornis

Hermissenda is one preparation that has contributed to an understanding of Pavlovian classical conditioning of visual (CS) and vestibular (US) turbulence at the cellular and molecular levels [60–63]. This animal can learn the sequential event of light and vestibular turbulence. Naive *Hermissenda* exhibit positive phototoxic behavior; that is, they move forward in response to light and contract their foot in response to turbulence—whereas after paired presentations of light and orbital rotation they hesitate to move toward light. Animals receiving the same amount of stimuli without overlapping the CS and US (i.e., pseudo-random conditioning) or light and/or rotation alone do not show any conditioned behavior in response to the light stimulus. This demonstrates that only the paired presentation of the CS and US results in associative learning. The conditioned behavior is evaluated by observing the foot length; naive animals extend their foot, whereas conditioned animals contract their foot in response to the light stimulus.

The central nervous system in *Hermissenda* is relatively simple, making it possible to identify the neurons in the neuronal circuits that are involved in the conditioning. The two sensory systems mediating the CS and US have been described in detail by Alkon and colleagues [64–68]. In addition, the convergent site providing the synaptic interaction between the CS and US has been identified [66, 69–72]. The CS is detected by a pair of eyes comprising five photoreceptor cells, which are subdivided into two types in terms of sensitivity to light: two type A photoreceptors and three type B photoreceptor cells. The vestibular sensing organ is a pair of statocysts comprising 13 hair cells. The caudal hair cells synapse onto the medial type B photoreceptor via a γ-aminobutyric acid (GABA)ergic synapse [73–75]. This synapse is the primary locus of plasticity after the conditioning. Because the medial type B photoreceptor is not only a photoreceptor but also a postsynaptic neuronal element, many studies have focused on this synapse to elucidate the mechanisms of classical conditioning from the viewpoint of biophysics, biochemistry, morphology, and molecular biology.

Two sensory receptors, photoreceptors and statocyst hair cells, mediating the CS and the US, respectively, are located in both the cerebral and pedal ganglia of the circumesophageal nervous system, and thus their synaptic projections remain intact even in the isolated brain preparation. Thus, this unique model system allows us to study in vitro conditioning employing the natural CS and US stimuli used for in vivo studies [76–78]. The mechanism of CS–US contiguity has been identified and has been the focus of neuroinformatic, biophysical, biochemical, morphologic, and molecular biologic analyses [79, 80]. Figure 3 shows a simplified circuit diagram involving the visuovestibular associative learning in *Hermissenda*.

Squire and Alvarez defined memory consolidation as "the molecular cascade and morphologic changes whereby synaptic modifications gradually become stable after learning" [81]. This definition also applies to invertebrate models of consolidated long-term memory (CLTM). The relation between the number of training events (TEs), defined by the number of paired presentations of the CS and US, and the number of retention days was evaluated from behavioral observations made by Epstein and colleagues showing the time window for the formation of STM to CLTM [82–84]. The STM lasts 7 min with one or two TEs. Nine TEs result in LTM within 60 min that lasts less than 1 day; and CLTM occurs 220 min after the conditioning and lasts for at least 6 days. STM requires no protein synthesis, whereas both LTM and CLTM require protein synthesis. Thus, the spectrum of events that correlate with the establishment of these memory stages and the transitions between them can be readily studied. The two-critical-periods theory of protein synthesis for LTM formation mentioned above has also been supported in this model system;

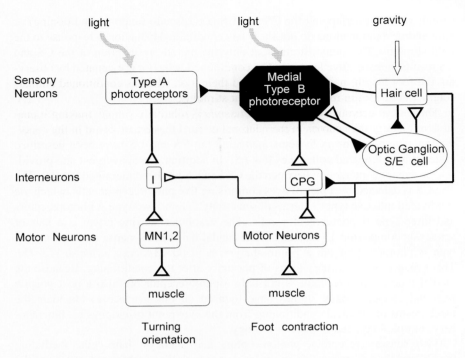

Fig. 3. Flow of visual and vestibular information in *Hermissenda crassicornis*. The conditioned stimulus is received at five photoreceptors, and the unconditioned stimulus is sensed by 13 hair cells. Photoreceptors are subdivided into two type A and three type B cells. The medial type B photoreceptor is not only a photosensitive neuron but also a postsynaptic neuron of statocyst hair cells. Thus, the primary conditioning effect is observed at the type B photoreceptor. Increased excitability of type B cells facilitates foot contraction. *Filled triangles*, inhibitory synapses; *open triangles*, excitatory synapses. *MN*, motor neuron; *CPG*, cerebropleural ganglion

the first critical period is up to 15min, and the other is 60–220min after the conditioning. The translation inhibitor anisomycin blocks memory recall when applied 13min and 60–220min after the conditioning. There is good memory recall from 10 to 60min. It is assumed that the period from 15 to 60min after the conditioning involves the activities of the proteins already made, but with no new protein synthesis [83].

Consistent with this result, morphologic modifications are observed at the axon terminal of the type B photoreceptor, which is the postsynaptic element for the vestibular hair cells and presumed to control whole-animal movement during phototaxis [78, 85]. Kawai et al. demonstrated that the entire volume of the terminal branch arborization of the type B photoreceptor axon starts to decrease significantly soon after several presentations of the CS and US, peaks at 10min, and remains at the same elevated level for up to 60min in an isolated preparation. This decrease was originally discovered in an in vivo preparation after acquisition of learning by

Alkon et al., who called it *focusing* [86]. During focusing, an increase in input resistance also occurs in the neurons of the in vitro preparation. The dynamic nature of the morphology and physiology is completely parallel for up to 60 min after the conditioning. Kawai et al. also demonstrated that no morphologic modification occurs in a pre-anisomycin-treated preparation [85]. The in vitro conditioning-induced synaptic focusing is also prevented by injecting the Ca^{2+} chelator BAPTA, the ryanodine receptor blocker dantrolene, or micromolar concentrations of ryanodine into the type B photoreceptor. These results indicate that morphologic modification after in vitro conditioning involves the ryanodine receptor [87].

The early stage of protein synthesis without mRNA modification is obvious in this preparation. The possible activities include interactions among existing substances in their various intracellular compartments. During the 60- to 220-min period, either anisomycin or actinomycin-D suppressed recall. This period is assumed to be the second period of protein synthesis. Cell adhesion molecules (CAMs) such as HNK-1, neural CAMs, and integrins have pivotal roles in long-term potentiation in both invertebrates and mammals [87–89]. The most common extracellular matrix protein tripeptide sequence, Arg-Gly-Asp (RGD), acts as a CAM-competitive inhibitor and can disrupt memory function. The latter period, when new dendritic spines and synapses are formed, is the time when CAM inhibitors are effective. Animals treated with RGD applied 10 min following conditioning exhibited complete inhibition of learning or no recall of the conditioned behavior. Animals treated with RGD applied 20–50 min after training, however, had a marked decline in inhibition [82]. LTM in *Hermissenda*, as well as transcription and translation inhibitor response criteria, is established within 60 min [83]. LTM consolidation requires 60–230 min [82, 83]. This interval is too long to be due to transmitting signals through synapses or new protein synthesis transported anterogradely from the soma.

Studies by Alkon et al. demonstrated the protein synthesis mechanism that is required for consolidating associative learning into LTM [90]. The application of a potent PKC activator, bryostatin, for 2 days before conditioning induces the synthesis of proteins that are necessary and sufficient for subsequent CLTM. Under normal conditions, two TEs with paired CS and US cause STM lasting 7 min; after a 4-h exposure to subnanomolar (0.1–0.25 ng/ml) concentrations of bryostatin on the 2 days preceding conditioning, however, the same two TEs produce CLTM that lasts longer than 1 week and is not blocked by anisomycin. Anisomycin, however, eliminates LTM lasting at least 1 week after nine TEs. Both the nine TEs alone and the two TEs with bryostatin exposure induce a comparable increase in the PKC α-isozyme substrate calexcitin in the type B photoreceptor, which shows a Pavlovian conditioning-dependent increase in phosphorylation and absolute quantity [91], and enhances PKC activity in the membrane fraction. The specific PKC antagonist Ro-32-0432 or anisomycin blocks bryostatin-induced protein synthesis as well as bryostatin-induced enhancement of behavioral conditioning [90]. Electrophysiologic measures of input resistance and long-lasting depolarization in response to a light stimulus also demonstrated that bryostatin induces excitability in the type B photoreceptor [92]. Bryostatin increases the synthesis of calexcitin in the type B photoreceptor, as occurs with Pavlovian classical conditioning of *Hermissenda* [93]. Bryostatin in low doses (0.1–0.25 ng/ml) initially enhances PKC activation followed by down-regulation and then prolonged enhancement of protein synthesis. Bryostatin-induced PKC enhancement of protein synthesis enhances the duration of the memory of Pavlovian classical conditioned responses. Bryostatin-induced PKC activation on days before training is sufficient to cause LTM. This LTM does not require protein synthesis after the training. PKC, after activation induced by bryostatin, is down-regulated by two distinct pathways: one that is proteasome-mediated and another that is mediated by phosphatases such as protein phosphatase 1 and protein phosphatase 2A [94]. Higher bryostatin concentrations (≤ 1.0 ng/ml) block memory retention because PKC synthesis cannot compensate for inactivation and down-regulation; therefore, the available PKC is depleted and memory retention is blocked [90].

Lymnaea can be conditioned with the same CS and US paired presentation as *Hermissenda*, but the underlying mechanisms for visuovestibular associative learning are suggested to be different [36, 95].

Conclusion

Cellular and molecular mechanisms underlying STM and LTM are reviewed based on observations of molluscan models. STM results from changes in the synaptic strength of preexisting neuronal connections that involve covalent modifications of preexisting proteins by various kinases. The synaptic plasticity underlying LTM is believed to involve protein synthesis and modulation of gene expression to induce new mRNA, protein synthesis, and morphologic modifications. These processes and their mechanisms in the three molluscan model systems explored herein likely have commonalities with those of mammals.

References

1. Pavolv I (1927) Conditioned reflexes. Oxford University Press, London.
2. Hebb DO (1949) The organization of behavior. Wiley, New York.
3. Sutton MA, Masters SE, Bagnall MW, et al (2001) Molecular mechanisms underlying a unique intermediate phase of memory in *Aplysia*. Neuron 31:143–154.
4. Sangha S, Scheibenstock A, McComb C, et al (2003) Intermediate and long-term memories of associative learning are differentially affected by transcription versus translation blockers in *Lymnaea*. J Exp Biol 206:1605–1613.
5. Carew TJ, Castellucci VF, Kandel ER (1971) An analysis of dishabituation and sensitization of the gill-withdrawal reflex in *Aplysia*. Int J Neurosci 2:79–98.
6. Carew TJ, Walters ET, Kandel ER (1981) Classical conditioning in a simple withdrawal reflex in *Aplysia californica*. J Neurosci 1:1426–1437.
7. Kandel ER (2001) The molecular biology of memory storage: a dialogue between genes and synapses. Science 294:1030–1038.

8. Castellucci V, Pinsker H, Kupfermann I, et al (1970) Neuronal mechanisms of habituation and dishabituation of the gill-withdrawal reflex in *Aplysia*. Science 167:1745–1748.
9. Castellucci VF, Frost WN, Goelet P, et al (1986) Cell and molecular analysis of long-term sensitization in *Aplysia*. J Physiol (Paris) 81:349–357.
10. Byrne JH (2003) Learning and memory: basic mechanisms. In: Squire LR, Bloom FE, Roberts JL, et al (eds) Fundamental neuroscience. Academic, San Diego, pp 1275–1298.
11. Glanzman DL (2006) The cellular mechanisms of learning in *Aplysia*: of blind men and elephants. Biol Bull 210:271–279.
12. Jessell TM, Kandel ER (1993) Synaptic transmission: a bidirectional and self-modifiable form of cell-cell communication. Cell 72(suppl):1–30.
13. Hawkins RD, Kandel ER, Bailey CH (2006) Molecular mechanisms of memory storage in *Aplysia*. Biol Bull 210:174–191.
14. Glanzman DL, Mackey SL, Hawkins RD, et al (1989) Depletion of serotonin in the nervous system of *Aplysia* reduces the behavioral enhancement of gill withdrawal as well as the heterosynaptic facilitation produced by tail shock. J Neurosci 9:4200–4213.
15. Montarolo PG, Goelet P, Castellucci VF, et al (1986) A critical period for macromolecular synthesis in long-term heterosynaptic facilitation in *Aplysia*. Science 234:1249–1254.
16. Bao JX, Kandel ER, Hawkins RD (1998) Involvement of presynaptic and postsynaptic mechanisms in a cellular analog of classical conditioning at *Aplysia* sensory-motor neuron synapses in isolated cell culture. J Neurosci 18:458–466.
17. Barbas D, DesGroseillers L, Castellucci VF, et al (2003) Multiple serotonergic mechanisms contributing to sensitization in *Aplysia*: evidence of diverse serotonin receptor subtypes. Learn Mem 10:373–386.
18. Braha O, Edmonds B, Sacktor T, et al (1993) The contributions of protein kinase A and protein kinase C to the actions of 5-HT on the L-type Ca^{2+} current of the sensory neurons in *Aplysia*. J Neurosci 13:1839–1851.
19. Chain DG, Casadio A, Schacher S, et al (1999) Mechanisms for generating the autonomous cAMP-dependent protein kinase required for long-term facilitation in *Aplysia*. Neuron 22:147–156.
20. Sutton MA, Carew TJ (2000) Parallel molecular pathways mediate expression of distinct forms of intermediate-term facilitation at tail sensory-motor synapses in *Aplysia*. Neuron 26:219–231.
21. Nakanishi K, Zhang F, Baxter DA, et al (1997) Role of calcium-calmodulin-dependent protein kinase II in modulation of sensorimotor synapses in *Aplysia*. J Neurophysiol 78:409–416.
22. Li Q, Roberts AC, Glanzman DL (2005) Synaptic facilitation and behavioral dishabituation in *Aplysia*: dependence on release of Ca^{2+} from postsynaptic intracellular stores, postsynaptic exocytosis, and modulation of postsynaptic AMPA receptor efficacy. J Neurosci 25:5623–5637.
23. Jin I, Hawkins RD (2003) Presynaptic and postsynaptic mechanisms of a novel form of homosynaptic potentiation at *Aplysia* sensory-motor neuron synapses. J Neurosci 23:7288–7297.
24. Bao JX, Kandel ER, Hawkins RD (1997) Involvement of pre- and postsynaptic mechanisms in posttetanic potentiation at *Aplysia* synapses. Science 275:969–973.
25. Antonov I, Antonova I, Kandel ER, et al (2001) The contribution of activity-dependent synaptic plasticity to classical conditioning in *Aplysia*. J Neurosci 21:6413–6422.
26. Sharma SK, Bagnall MW, Sutton MA et al (2003) Inhibition of calcineurin facilitates the induction of memory for sensitization in *Aplysia*: requirement of mitogen-activated protein kinase. Proc Natl Acad Sci USA 100:4861–4866.
27. Dash PK, Hochner B, Kandel ER (1990) Injection of the cAMP-responsive element into the nucleus of *Aplysia* sensory neurons blocks long-term facilitation. Nature 345:718–721.
28. Bailey CH, Chen M (1988) Long-term sensitization in *Aplysia* increases the number of presynaptic contacts onto the identified gill motor neuron L7. Proc Natl Acad Sci USA 85:9356–9359.

29. Bailey CH, Chen M (1988) Long-term memory in *Aplysia* modulates the total number of varicosities of single identified sensory neurons. Proc Natl Acad Sci USA 85:2373–2377.
30. Bailey CH, Chen M (1989) Structural plasticity at identified synapses during long-term memory in *Aplysia*. J Neurobiol 20:356–372.
31. Kim JH, Udo H, Li HL, et al (2003) Presynaptic activation of silent synapses and growth of new synapses contribute to intermediate and long-term facilitation in *Aplysia*. Neuron 40:151–165.
32. Audesirk TE, Alexander JE Jr, Audesirk GJ, et al (1982) Rapid, nonaversive conditioning in a freshwater gastropod. I. Effects of age and motivation. Behav Neural Biol 36:379–390.
33. Kemenes G, Benjamin PR (1989) Appetitive learning in snails shows characteristics of conditioning in vertebrates. Brain Res 489:163–166.
34. Kojima S, Nanakamura H, Nagayama S, et al (1997) Enhancement of an inhibitory input to the feeding central pattern generator in *Lymnaea stagnalis* during conditioned taste-aversion learning. Neurosci Lett 230:179–182.
35. Kawai R, Sunada H, Horikoshi T, et al (2004) Conditioned taste aversion with sucrose and tactile stimuli in the pond snail *Lymnaea stagnalis*. Neurobiol Learn Mem 82:164–168.
36. Sakakibara M, Kawai R, Kobayashi S, et al (1998) Associative learning of visual and vestibular stimuli in *Lymnaea*. Neurobiol Learn Mem 69:1–12.
37. Lukowiak K, Cotter R, Westly J, et al (1998) Long-term memory of an operantly conditioned respiratory behaviour pattern in *Lymnaea stagnalis*. J Exp Biol 201(Pt 6):877–882.
38. Lukowiak K, Adatia N, Krygier D, et al (2000) Operant conditioning in *Lymnaea*: evidence for intermediate- and long-term memory. Learn Mem 7:140–150.
39. Kemenes G, Elliot CJH, Benjamin PR (1986) Chemical and tactile inputs to the *Lymnaea* feeding system: effects on behaviour and neural circuitry. J Exp Biol 122:113–137.
40. Benjamin PR, Staras K, Kemenes G (2000) A systems approach to the cellular analysis of associative learning in the pond snail *Lymnaea*. Learn Mem 7:124–131.
41. Hatakeyama D, Fujito Y, Sakakibara M, et al (2004) Expression and distribution of transcription factor CCAAT/enhancer-binding protein in the central nervous system of *Lymnaea stagnalis*. Cell Tissue Res 318:631–641.
42. Lukowiak K, Sangha S, McComb C, et al (2003) Associative learning and memory in *Lymnaea stagnalis*: how well do they remember? J Exp Biol 206:2097–2103.
43. Martens KR, De Caigny P, Parvez K, et al (2007) Stressful stimuli modulate memory formation in *Lymnaea stagnalis*. Neurobiol Learn Mem 87:391–403.
44. Kojima S, Kobayashi S, Yamanaka M, et al (1998) Sensory preconditioning for feeding response in the pond snail, *Lymnaea stagnalis*. Brain Res 808:113–115.
45. Fulton D, Kemenes I, Andrew RJ, et al (2005) A single time-window for protein synthesis-dependent long-term memory formation after one-trial appetitive conditioning. Eur J Neurosci 21:1347–1358.
46. Alexander J Jr, Audesirk TE, Audesirk GJ (1984) One-trial reward learning in the snail *Lymnea stagnalis*. J Neurobiol 15:67–72.
47. Kemenes I, Kemenes G, Andrew RJ, et al (2002) Critical time-window for NO-cGMP-dependent long-term memory formation after one-trial appetitive conditioning. J Neurosci 22:1414–1425.
48. Kemenes G, Kemenes I, Michel M, et al (2006) Phase-dependent molecular requirements for memory reconsolidation: differential roles for protein synthesis and protein kinase A activity. J Neurosci 26:6298–6302.
49. Sugai R, Azami S, Shiga H, et al (2007) One-trial conditioned taste aversion in *Lymnaea*: good and poor performers in long-term memory acquisition. J Exp Biol 210:1225–1237.
50. Straub VA, Styles BJ, Ireland JS, et al (2004) Central localization of plasticity involved in appetitive conditioning in *Lymnaea*. Learn Mem 11:787–793.
51. Sekiguchi T, Yamada A, Suzuki H (1997) Reactivation-dependent changes in memory states in the terrestrial slug *Limax flavus*. Learn Mem 4:356–364.

52. Sangha S, Morrow R, Smyth K, et al (2003) Cooling blocks ITM and LTM formation and preserves memory. Neurobiol Learn Mem 80:130–139.
53. Sangha S, Scheibenstock A, Morrow R, et al (2003) Extinction requires new RNA and protein synthesis and the soma of the cell right pedal dorsal 1 in *Lymnaea stagnalis*. J Neurosci 23:9842–9851.
54. Alberini CM, Ghirardi M, Metz R, et al (1994) C/EBP is an immediate-early gene required for the consolidation of long-term facilitation in *Aplysia*. Cell 76:1099–1114.
55. Taubenfeld SM, Milekic MH, Monti B, et al (2001) The consolidation of new but not reactivated memory requires hippocampal C/EBPbeta. Nat Neurosci 4:813–818.
56. Guzowski JF, Setlow B, Wagner EK, et al (2001) Experience-dependent gene expression in the rat hippocampus after spatial learning: a comparison of the immediate-early genes Arc, c-fos, and zif268. J Neurosci 21:5089–5098.
57. Anokhin KV, Rose SP (1991) Learning-induced increase of immediate early gene messenger RNA in the chick forebrain. Eur J Neurosci 3:162–167.
58. Hatakeyama D, Sadamoto H, Ito E (2004) Real-time quantitative RT-PCR method for estimation of mRNA level of CCAAT/enhancer binding protein in the central nervous system of *Lymnaea stagnalis*. Acta Biol Hung 55:157–161.
59. Hatakeyama D, Sadamoto H, Watanabe T, et al (2006) Requirement of new protein synthesis of a transcription factor for memory consolidation: paradoxical changes in mRNA and protein levels of C/EBP. J Mol Biol 356:569–577.
60. Alkon DL (1983) Learning in a marine snail. Sci Am 249:70–74, 76–78, 80–84.
61. Crow TJ, Alkon DL (1978) Retention of an associative behavioral change in *Hermissenda*. Science 201:1239–1241.
62. Alkon DL (1989) Simple systems. In: Memory traces in the brain. Cambridge University Press, Cambridge, pp 13–28.
63. Alkon DL, Nelson TJ, Zhao W, et al (1998) Time domains of neuronal Ca^{2+} signaling and associative memory: steps through a calexcitin, ryanodine receptor, K^+ channel cascade. Trends Neurosci 21:529–537.
64. Alkon DL, Fuortes MG (1972) Response of photoreceptors in *Hermissenda*. J Gen Physiol 60:631–649.
65. Alkon DL (1973) Neural organization of a molluscan visual system. J Gen Physiol 61:444–461.
66. Alkon DL (1973) Intersensory interactions in *Hermissenda*. J Gen Physiol 62:185–202.
67. Alkon DL, Bak A (1973) Hair cell generator potentials. J Gen Physiol 61:619–637.
68. Detwiler PB, Alkon DL (1973) Hair cell interactions in the statocyst of *Hermissenda*. J Gen Physiol 62:618–642.
69. Alkon DL, Akaike T, Harrigan J (1978) Interaction of chemosensory, visual, and statocyst pathways in *Hermissenda crassicornis*. J Gen Physiol 71:177–194.
70. Akaike T, Alkon DL (1980) Sensory convergence on central visual neurons in *Hermissenda*. J Neurophysiol 44:501–513
71. Crow T, Tian LM (2000) Monosynaptic connections between identified A and B photoreceptors and interneurons in *Hermissenda*: evidence for labeled-lines. J Neurophysiol 84:367–375.
72. Crow T, Tian LM (2004) Statocyst hair cell activation of identified interneurons and foot contraction motor neurons in *Hermissenda*. J Neurophysiol 91:2874–2883.
73. Tabata M, Alkon DL (1982) Positive synaptic feedback in visual system of nudibranch mollusk *Hermissenda crassicornis*. J Neurophysiol 48:174–191.
74. Alkon DL, Sanchez-Andres JV, Ito E, et al (1992) Long-term transformation of an inhibitory into an excitatory GABAergic synaptic response. Proc Natl Acad Sci USA 89:11862–11866.
75. Collin C, Ito E, Oka K, et al (1992) The role of calcium in prolonged modification of a GABAergic synapse. J Physiol (Paris) 86:139–145.
76. Matzel LD, Lederhendler, II, Alkon DL (1990) Regulation of short-term associative memory by calcium-dependent protein kinase. J Neurosci 10:2300–2307.

77. Matzel LD, Muzzio IA, Talk AC (1996) Variations in learning reflect individual differences in sensory function and synaptic integration. Behav Neurosci 110:1084–1095.
78. Kawai R, Horikoshi T, Yasuoka T, et al (2002) In vitro conditioning induces morphological changes in *Hermissenda* type B photoreceptor. Neurosci Res 43:363–372.
79. Crow T, Tian LM (2006) Pavlovian conditioning in *Hermissenda*: a circuit analysis. Biol Bull 210:289–297.
80. Blackwell KT (2006) Subcellular, cellular, and circuit mechanisms underlying classical conditioning in *Hermissenda crassicornis*. Anat Rec B New Anat 289:25–37.
81. Squire LR, Alvarez P (1998) Retrograde amneasia and memory consolidation: a neurobiological perspective. In: Squire LR, Kosslyn L (eds) Findings and current opinion in cognitive neuroscience. MIT Press, Cambridge, MA, pp 75–84.
82. Epstein DA, Epstein HT, Child FM, et al (2000) Memory consolidation in *Hermissenda crassicornis*. Biol Bull 199:182–183.
83. Epstein HT, Child FM, Kuzirian AM, et al (2003) Time windows for effects of protein synthesis inhibitors on Pavlovian conditioning in *Hermissenda*: behavioral aspects. Neurobiol Learn Mem 79:127–131.
84. Epstein HT, Kuzirian AM, Child FM, et al (2004) Two different biological configurations for long-term memory. Neurobiol Learn Mem 81:12–18.
85. Kawai R, Yasuoka T, Sakakibara M (2003) Dynamical aspects of in vitro conditioning in *Hermissenda* type B photoreceptor. Zoolog Sci 20:1–6.
86. Alkon DL, Ikeno H, Dworkin J, et al (1990) Contraction of neuronal branching volume: an anatomic correlate of Pavlovian conditioning. Proc Natl Acad Sci USA 87:1611–1614.
87. Kawai R, Horikoshi T, Sakakibara M (2004) Involvement of the ryanodine receptor in morphologic modification of *Hermissenda* type B photoreceptors after in vitro conditioning. J Neurophysiol 91:728–735.
88. Pinkstaff JK, Lynch G, Gall CM (1998) Localization and seizure-regulation of integrin beta 1 mRNA in adult rat brain. Brain Res Mol Brain Res 55:265–276.
89. Wildering WC, Hermann PM, Bulloch AG (1998) Neurite outgrowth, RGD-dependent, and RGD-independent adhesion of identified molluscan motoneurons on selected substrates. J Neurobiol 35:37–52.
90. Alkon DL, Epstein H, Kuzirian A, et al (2005) Protein synthesis required for long-term memory is induced by PKC activation on days before associative learning. Proc Natl Acad Sci USA 102:16432–16437.
91. Kuzirian AM, Epstein HT, Buck D, et al (2001) Pavlovian conditioning-specific increases of the Ca^{2+}- and GTP-binding protein, calexcitin in identified *Hermissenda* visual cells. J Neurocytol 30:993–1008.
92. Kuzirian AM, Epstein HT, Gagliardi CJ, et al (2006) Bryostatin enhancement of memory in *Hermissenda*. Biol Bull 210:201–214.
93. Nelson TJ, Quattrone A, Kim J, et al (2003) Calcium-regulated GTPase activity in the calcium-binding protein calexcitin. Comp Biochem Physiol B Biochem Mol Biol 135: 627–638.
94. Morrione A, Romano G, Navarro M, et al (2000) Insulin-like growth factor I receptor signaling in differentiation of neuronal H19-7 cells. Cancer Res 60:2263–2272.
95. Sakakibara M (2006) Comparative study of visuo-vestibular conditioning in *Lymnaea stagnalis*. Biol Bull 210:298–307.

Role of the Noradrenergic System in Synaptic Plasticity in the Hippocampus

Ming-Yuan Min[1], Hsiu-Wen Yang[2], and Yi-Wen Lin[3]

Summary

The cortical noradrenergic (NAergic) system, which originates from the locus coeruleus (LC) located in the pons, plays an important role in cortical plasticity and many other brain functions. In rats in which the NAergic system has been eliminated by 6-hydroxydopamine during the neonatal period, induction of long-term potentiation (LTP) at CA1 synapses in the hippocampus is impaired, whereas induction of long-term depression is unaffected. Bath application of norepinephrine, a β-adrenergic receptor agonist, or activators of effector molecules downstream of the β-adrenergic receptor restores LTP. Similarly, activation of β-adrenergic receptors enhances associative LTP induced by paired stimuli to two independent synaptic inputs on the same postsynaptic neuron. The time window within which LTP can be induced by paired stimuli is increased by β-adrenergic receptor activation, but the magnitude of LTP is not affected. The signaling molecules involved in enhancement of the homosynaptic and associative LTP following β-adrenergic receptor activation are the same and include protein kinase A and mitogen-activated protein kinases. These experimental results suggest that a simultaneous increase in the activity of LC neurons during induction protocols may have a permissive role in the induction of homosynaptic and associative LTP in the hippocampus.

Key words LTP, LTD, β-Adrenergic, 6-Hydroxydopamine, Locus coeruleus

---

[1]Department of Life Science & Institute of Zoology, College of Life Science, National Taiwan University, No. 1, Sec. 4, Roosevelt Road, Taipei 106, Taiwan

[2]Department of Biomedical Science, Chung-Shan Medical University, No. 110, Sec. 1, Chien-Kuo N. Road, Taichung 402, Taiwan

[3]Institute of Biomedical Science, Academia Sinica, No. 128, Sec. 2, Academia Road, Taipei 115, Taiwan

Introduction

Long-term potentiation (LTP) and long-term depression (LTD) are use-dependent changes in synaptic efficacy that are generally accepted as the cellular substrate for memory storage in the brain [1, 2]. They have attracted much attention during the past few decades, and the mechanisms involved in their induction and expression have been well characterized [2–4]. Activation of the *N*-methyl-D-aspartate (NMDA) subtype of glutamate receptors is required for induction of both LTP and LTD. The voltage-dependent nature of the NMDA receptor makes it an ideal detector for the total postsynaptic response produced by a population of activated synapses. These active synapses are potentiated when the total postsynaptic response they produce exceeds a critical value, referred to as the modification threshold (θm), and are depressed if the total response is greater than zero but less than θm [5, 6].

This idea of regulation of synaptic strength is called the BCM theory, named after the three scientists, Bienenstock, Cooper, and Murp, who proposed it [5]. An example of BCM modification for synaptic efficacy (or BCM curve) at the CA1 synapse is shown in Fig. 1. According to this theory, a total postsynaptic response greater than θm activates large numbers of NMDA receptors, which leads to LTP by the activation of protein kinases, which phosphorylate AMPA receptors. In contrast, a total postsynaptic response greater than zero but less than θm activates fewer NMDA receptors, which leads to LTD because of the activation of phosphatases and the subsequent dephosphorylation of AMPA receptors [4, 7].

Norepinephrine (NE), one of the most important neuronal modulators in the brain, is involved in the regulation of many brain functions, including the wake/

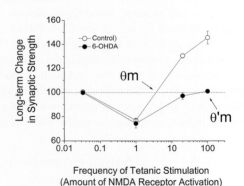

Fig. 1. Effect of norepinephrine (NE) on the BCM curve. The BCM curve shows the relation between the frequency of tetanic stimulation [or the amount of *N*-methyl-D-aspartate (*NMDA*) receptor activation by tetanic stimulation] and the resultant long-term change in synaptic strength in control slices (*open circles*). The *arrow* indicates the modification threshold (θ*m*). Note the right shift of θm in the control slices to θ′m in the 6-hydroxydopamine (*6-OHDA*)-treated slices (*filled circles*), suggesting that NE depletion impairs long-term potentiation (LTP) but has no effect on long-term depression (LTD) induction. (From Yang et al. [12], with permission)

sleep cycle [8], memory storage [9], synaptic/cortical plasticity [10–12], autonomic functions [13], and pain modulation [14]. In the neocortex and hippocampus, NE fibers mainly originate from the locus coeruleus (LC), located in the pons. In vitro studies have suggested that NE, acting on β-adrenergic receptors, has a significant effect on synaptic plasticity at CA1 synapses [10, 12, 15, 16] and mossy fiber synapses [17]. In this chapter, we first discuss the permissive role of NE in modulating synaptic plasticity based on previous studies, followed by some interesting issues that need to be examined further.

Role of the Noradrenergic System in Homosynaptic Plasticity

At hippocampal CA1 synapses, bath application of 10 μM NE during the delivery of conditioned stimulation to induce a long-term change in synaptic strength results in a left shift of θm in the BCM curve. That is, application of NE does not have any effect on the magnitude of LTP induced by high-frequency stimulation at 50 or 100 Hz but blocks induction of LTD by low-frequency stimulation (1 Hz) and enhances the effect of 10-Hz stimulation—which alone does not result in any significant change in synaptic strength in control conditions—to induce LTP [15]. Similar observations have been made at mossy fiber synapses on CA3 pyramidal neurons [17, 18]. An in vivo study also showed that induction of LTP by mild tetanus stimulation is impaired in animals in which endogenous catecholamine was depleted by injecting 6-hydroxyl dopamine (6-OHDA), but LTP can still be induced by strong tetanus [19]. These observations therefore suggest that the cortical noradrenergic (NAergic) system might have a permissive role in modulating LTP induction in the hippocampus.

Nevertheless, a critical question remains: How do conditioned stimuli for LTP induction activate the NAergic pathway to enhance LTP induction? A high density of NAergic fibers is found in the stratum radiatum of the CA1 region (Fig. 2) and the stratum lucidum of the CA3 region [20, 21]. The overlapping distribution of NAergic and glutamatergic fibers of the Shaffer collateral branches provides an opportunity for simultaneous activation of glutamatergic and NAergic fibers during tetanus. Consistent with this argument, studies in which electrophysiological and neurochemical data were simultaneously recorded in the hippocampus in vivo showed that the local NE concentration can rise to several times the basal value during tetanus for LTP induction [22, 23]. Another potential approach to addressing this question is to examine synaptic plasticity in hippocampal slices from animals in which endogenous NAergic fibers in the hippocampus have been depleted. The treatment of neonatal rats with 6-OHDA might provide a model meeting this requirement, as it results in a persistent loss of catecholaminergic fibers in the cerebral cortex but not in subcortical areas (Fig. 2) [24]. Using hippocampal slices from 6-OHDA-treated rats, we found that the LTP induced by theta burst stimulation decays within 15 min (Fig. 3A,C), whereas the LTD induced by 900 pulses at 1 Hz is not affected (Fig. 3B,C) [12]. Taking these results together, it reveals a right

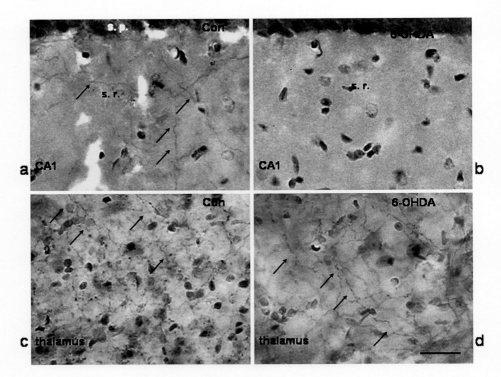

Fig. 2. Animal model in which endogenous NE is depleted by 6-OHDA. Dopamine-β-hydroxylase (DBH) immunohistochemistry performed in the stratum radiatum (*s.r.*) of the hippocampal CA1 area in control (*Con*) (**a**) and in 6-OHDA-treated animals (**b**). Thalamus area in the same section in the control animal (**c**) and the 6-OHDA-treated animal (**d**). Note the DBH-immunoreactive fibers (indicated by *arrows*) in all photographs except **b**. *s.p.*, stratum pyramidale. *Bars* 50 μm. (From Yang et al. [12], with permission)

shift of θm in the BCM curve that is consistent with the effect of perfusion slices with 10 μM NE [15]. Interestingly, a similar observation has been reported at mossy fiber synapses, where the LTP induced by mild tetanus decays within 15 min when β-adrenergic receptors are blocked by timolol [17]. Bath application of NE restores the expression of LTP to normal and blocks LTD induction (Fig. 3A,B) in slices from 6-OHDA-treated animals but has no effect on the magnitude of LTP in slices from control animals (Fig. 3C) [12]. These observations suggest that NAergic fibers are recruited to enhance LTP induction in normal hippocampal slices, whereas a higher tetanus intensity is required for LTP induction in slices lacking NAergic innervation. Similar conclusions were drawn in a recent study that examined the expression of LTP in several strains of mice with various levels of endogenous NE [16].

The enhancement of LTP and blockage of LTD by NE in the hippocampus appears to occur through the activation of β-adrenergic receptors [12, 15–17]. In

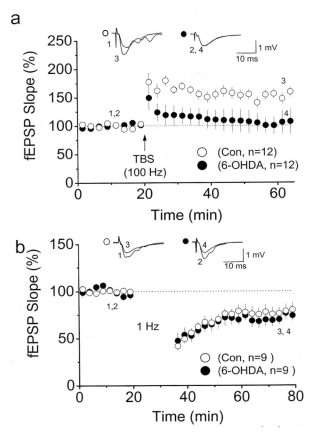

Fig. 3. LTP and LTD in slices from 6-OHDA-lesioned animals. **a** LTP at CA1 synapses induced in slices of control (*open circles*) and 6-OHDA-treated (*filled circles*) rats using theta burst stimulation (*TBS*), which consists of 10 bursts at 5 Hz with each burst consisting of four pulses at 100 Hz. **b** LTD induced by 900 pulses at 1 Hz stimulation. Note the rapid decay in the LTP, but not the LTD, induced in slices from 6-OHDA-treated rats. (From Yang et al. [12], with permission)

control slices, the effect of NE on the BCM curve is blocked by timolol, a selective β-adrenergic receptor antagonist, but not by phentolamine, a selective α-adrenergic receptor [15]. Similarly, isoproterenol, a selective β-adrenergic receptor agonist, but not phenylephrine, a selective α-adrenergic receptor agonist, restores LTP to normal in slices from 6-OHDA-treated animals (Fig. 4) [12] and from mice with low endogenous NE levels [16]. Interestingly, the blockage of depotentiation (low frequency-induced LTD after LTP induction) at CA1 synapses requires the activation of both α- and β-adrenergic receptors [15], suggesting that the mechanisms underlying LTD induction and depotentiation are different [25].

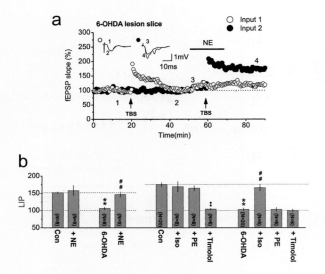

Fig. 4. Activation of β-adrenergic, but not α-adrenergic, receptors restores LTP in slices from 6-OHDA-treated rats. **a** A typical experiment on a slice from a 6-OHDA-treated rat in which two independent Schaffer collateral branches were stimulated. Theta burst stimulation (*TBS*) was applied to one of the two inputs (input 1) to induce LTP in normal conditions; 40 min later, TBS was applied to the other input (input 2) to induce LTP with simultaneous application of NE during delivery of TBS. The resultant LTPs are compared. **b** Summarized results show that application of NE restores LTP in slices from 6-OHDA-treated rats and the summarized results of pharmacological experiments. Note that in slices from control rats LTP induction is not affected by isoproterenol (*Iso*), a β-adrenergic receptor agonist, or phenylephrine (*PE*), an α-adrenergic receptor agonist; it is blocked by timolol, a β-adrenergic receptor antagonist. Also note in slices from 6-OHDA-treated rats LTP is restored by application of Iso but not PE, and this effect is prevented when the β-adrenergic receptor is blocked by timolol. (From Yang et al. [12], with permission)

Role of the NAergic System in Associative Plasticity

A significant weakness in suggesting a physiological role of the conventional LTP and LTD induced by tetanus or prolonged low-frequency stimulation has been the question of whether these stimulating paradigms realistically resemble any physiological function in the brain. Recently, it was demonstrated that LTP/LTD-like changes in synaptic strength can also be induced by the simultaneous spiking of a pair of pre- and postsynaptic neurons within a precise time window, so-called spike timing-dependent plasticity (STDP) [26–28]. Obviously, this form of synaptic plasticity fits very well with the Hebbian learning rules that: (1) neurons that fire in synchrony become wired together (i.e., when the presynaptic axon is active and the postsynaptic neuron simultaneously is strongly activated, the synapse formed by the presynaptic axon is strengthened); and (2) neurons that fire out of synchrony lose their link (see also Chapter 25 in [29]). It is now generally accepted that STDP provides a more genuine cellular model for experience-driven change in brain function or the Hebbian learning rules than does conventional LTP/LTD [26–28, 30, 31].

STDP is bidirectional; that is, synaptic efficacy can be either potentiated or depressed by paired pre- and postsynaptic spiking, depending on both the timing interval and the temporal order of the pre- and postsynaptic spiking. Generally speaking, to induce STDP, the timing interval between paired pre- and postsynaptic spiking has to be less than ~25 ms, whereas a significantly wider time window (up to ~100 ms) has been suggested for LTD induction [32, 33]. As for the temporal order of pre- and postsynaptic spiking, repeated paired pre/postsynaptic spiking results in LTP if presynaptic stimulation precedes postsynaptic stimulation, whereas it results in LTD if the temporal order of pre/postsynaptic spiking is reversed (Fig. 5).

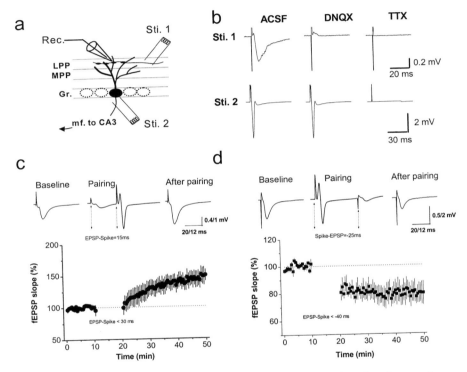

Fig. 5. Spike-timing-dependent plasticity (STDP) at a synapse of the lateral perforant path on a granule cell in the dentate gyrus. **a** The arrangement of the recording electrode (*Rec.*) and stimulating electrodes (*Sti. 1, Sti. 2*). **b** Evoked neuronal activity by *Sti. 1* is field EPSP (*fEPSP*) activity, as it is completely blocked by the AMPA receptor antagonist DNQX. The activity evoked by *Sti. 2* is the field somatic spike (fSS), as it is insensitive to DNQX, but is completely blocked by tetrodotoxin (*TTX*). **c** LTP induction by paired fEPSP–fSS stimulation. The *upper traces* show baseline fEPSP activity (*left*), fEPSP and fSS during pairing (*middle*), and fEPSP activity after pairing (*right*) for one experiment. Note the potentiation of fEPSP activity after paired fEPSP-afSS stimulation with $\Delta t = 15$ ms. The *lower plot* shows the summarized results for nine experiments, in which the Δt for the paired fEPSP–afSS stimulation was <30 ms. **d** LTD induction by paired fSS–fEPSP stimulation. Note the depression of fEPSP activity (compare the left and right **insets**) after paired afSS–fEPSP stimulation with $\Delta t = -25$ ms (see middle **inset**). The lower plot shows the summarized results for eight experiments, in which Δt for the paired fEPSP–afSS stimulation was <|–40|ms. (From Lin et al. [25], with permission)

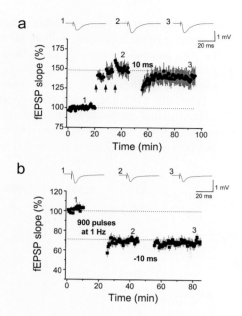

Fig. 6. Induction of STDP is occluded by saturated homosynaptic LTP/LTD. **a** Homosynaptic LTP induced by three trains of 100 pulses at 100 Hz (intertrain interval 30 s), repeated three times at intervals of 5 min (*arrows*); following this homosynaptic LTP, induction of LTP by paired stimulation with an interval of 10 ms is occluded. **b** LTD induced by stimulation with 900 pulses at 1 Hz; following this homosynaptic LTD, induction of LTD by paired stimulation with an interval of −10 ms is again occluded. (From Lin et al. [25], with permission)

STDP-like LTP and LTD both require activation of NMDA receptors [10, 32, 33]. Again, the voltage-dependent nature of the NMDA receptor makes it an ideal detector for the correlated pre- and postsynaptic spiking activity. Depolarization of the membrane potential by the action potential back-propagated along dendrites of postsynaptic neurons removes magnesium, which blocks the pore site of the NMDA receptor in the resting condition and allows the NMDA receptor to be activated by glutamate released presynaptically during the paired pre- and postsynaptic spiking. We have recently characterized the cellular mechanisms underlying induction of STDP at a synapse of the lateral perforant pathway on a granule cell in the dentate gyrus (Fig. 5). We found that the signaling molecules involved in STDP induction are similar to those involved in conventional LTP and LTD because the saturated synaptic potentiation or depression caused by, respectively, tetanus or low-frequency stimulation, occludes the induction of STDP (Fig. 6) [25]. Similar to the homosynaptic plasticity discussed above, STDP is also subject to modification by the NAergic system. We have reported that by acting at β-adrenergic receptors NE enhances STDP by increasing the time window of the pre/postsynaptic activation required for LTP induction without changing the magnitude of LTP (Fig. 7) [10].

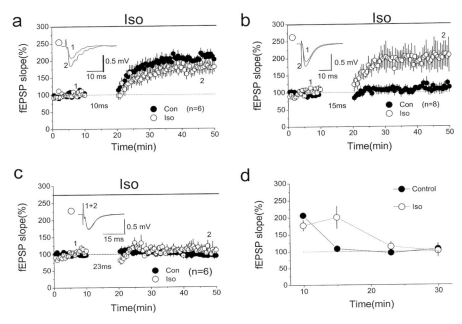

Fig. 7. Enhancement of STDP by activation of β-adrenergic receptors. **a** Application of 1 μM isoproterenol (*Iso*) has no significant effect on the magnitude of LTP induced using the pairing protocol with a 10-ms interval. **b** Application of Iso enhances LTP induction using the pairing protocol with a 15-ms interval, which does not induce LTP under control conditions. **c** Application of Iso has no significant effect when the interval is increased to 23 ms. **d** Summarized results for **a**–**c** and those using a 30-ms interval. In all of the plots, the *open circles* are results in the presence of Iso and the *filled circles* results from control experiments (*Con*). Note that there is no significant difference in the LTP induced in control and Iso experiments, except for that using a 15-ms interval. (From Lin et al. [10], with permission)

Molecular Signaling Cascades Activated to Enhance LTP

Activation of β-adrenergic receptors is known to increase the cytoplasmic concentration of cyclic adenosine monophosphate (cAMP), which in turn activates cAMP-dependent protein kinase A (PKA). This signaling pathway seems to act in parallel with the signaling pathway involved in LTP induction, which requires activation of calcium/calmodulin-dependent protein kinase II (CaMKII) by calcium influx through activated NMDA receptors during tetanus [4]. However, several studies have suggested that an increase in cAMP levels in postsynaptic neurons is required for LTP induction under control conditions [34–36]. Consistent with these observations, we found that bath application of activators of Gs protein or adenylyl cyclase (AC) also restored LTP in slices from 6-OHDA-treated animals (Fig. 8) [12]. The type I and VIII ACs, which are present at high amounts in neurons, are good candidate proteins for linking these two pathways. Unlike other AC subtypes, which

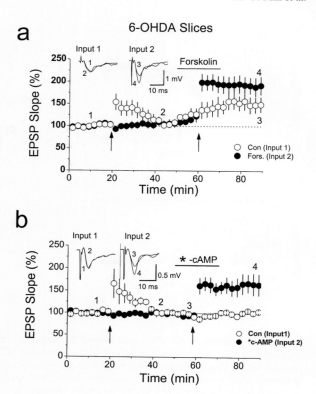

Fig. 8. Membrane-permeating cyclic adenosine monophosphate (*cAMP*) and protein kinase A (PKA) activators restore LTP in slices from 6-OHDA-treated rats. **a** LTP induction is restored by application of the membrane-permeating cAMP activator, forskolin, at Input 2. **b** Similar effect of the PKA activators 8-bromo-cAMP ($n = 4$) and dibutyryl-cAMP ($n = 4$) is also observed. The results of these two PKA activators are pooled. (From Yang et al. [12], with permission)

require an external signal for their activation, types I and VIII can be directly activated by CaMKII intracellularly and contribute to the cAMP increase in the cytoplasm [37]. Thus, the cAMP increase required for LTP induction might occur via activation of type I or VIII ACs by CaMKII [38, 39]. In hippocampal slices taken from type VIII AC knockout mice, induction of LTP is impaired, demonstrating the involvement of this protein in LTP induction [40].

It is likely that the permissive effect of the β-adrenergic receptor on homosynaptic LTP induction might be through the activation of other types of ACs downstream of the Gs proteins activated by β-adrenergic receptors, which in turn cause a sufficiently large increase in cytoplasmic cAMP for LTP induction. Therefore, cooperative activation of other types ACs by β-adrenergic receptors with activation of types I and VIII by NMDA receptor signaling leads to LTP induction. This could also explain why, in hippocampal slices that lack NAergic fiber innervation, strong tetanus is required for LTP induction, as more type I or VIII ACs must be activated to produce enough cAMP for LTP induction.

At hippocampal CA1 synapses, the expression of LTP is due to enhancement of AMPA receptor function, either by phosphorylation of existing AMPA receptors at synaptic sites or by recruitment of new AMPA receptors to the synaptic activation zone [41]. In addition to being directly involved in modulation of AMPA

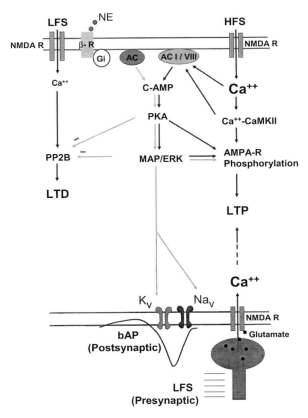

Fig. 9. Signaling cascades involved in the enhancing effect of β-adrenergic receptors (β-R) on synaptic plasticity at CA1 synapses. *Top right* High-frequency stimulation (*HFS*) activates many NMDA receptors (*NMDA R*) and causes a large amount of calcium influx and activation of CaMKII. The activated CaMKII either phosphorylates the AMPA receptor (*AMPA-R*) itself, leading to LTP, or increases cytoplasmic levels of cAMP by stimulating type I or VIII ACs (*ACI/VIII*), which activate PKA or the more downstream effectors MAP/ERK to cause phosphorylation of AMPA receptors and LTP induction (*black arrows*). *Middle* Activation of β-adrenergic receptors increases cytoplasmic levels of cAMP to enhance LTP induction, either by stimulating type I/VIII ACs or other types of AC (*gray arrows*). Activation of β-adrenergic receptors also inhibits phosphatase 2B (*PP2B*) and prevents induction of LTD by low-frequency stimulation (*LFS*) (*bottom*). As regards STDP, paired pre-/postsynaptic spiking causes activation of many NMDA receptors and leads to LTP as homosynaptic LTP. In addition, the activated PKA–MAP/ERK signaling pathway also modulates voltage-dependent potassium (K_v) and sodium (Na_v) channels function to shape the back-propagating action potential (*bAP*), which in turn changes the time window for LTP induction

receptor function, the signaling molecules activated following β-adrenergic receptor activation must also target other ion channels or receptors to enhance the associative form of LTP or STDP. This is because β-adrenergic receptor activation does not affect the magnitude of LTP but increases the time window of the pre/postsynaptic activation required for LTP induction (Fig. 9) [10]. For example, one possible candidate molecule is the Kv4.3 channel, which is responsible for a transient (or

A-type) potassium current reported to play an important role in controlling back-propagation of the action potential along dendrites in pyramidal neurons in the hippocampus [42]. Kv4.3 channel function is also subject to regulation by activation of β-adrenergic receptors through the phosphorylation of the Kv4.3 channels by PKA [42]. As a result, the time window of the pre/postsynaptic activation required for LTP induction is changed because of the changed properties of the back-propagated postsynaptic spike. Another possibility is that β-adrenergic receptors are located on the presynaptic terminals, and their activation results in a change in the profile of glutamate release presynaptically, which in turn changes the time window of pre/postsynaptic activation.

According to the BCM theory, during prolonged low-frequency stimulation fewer NMDA receptors are activated, there is less calcium influx into the postsynaptic cytoplasm, and phosphatase 2B (or calcineurin) is activated, rather than CaMKII, causing dephosphorylation of existing AMPA receptors and resulting in LTD [4]. Activation of the β-adrenergic receptor–cAMP–PKA pathway activates inhibitor I, which inhibits calcineurin (phosphatase 2B) and blocks LTD [4, 43, 44]. Interestingly, LTD induction is not affected in slices from 6-OHDA-treated animals [12], suggesting that NAergic fibers are not recruited during low-frequency stimulation, which may be due to their higher threshold for activation. Consistent with this argument, we found that STDP induction is also not affected [10], as presynaptic stimulation is given at low frequency during its induction.

Other Considerations

If recruitment of NAergic fibers occurs only during tetanus stimulation, which does not resemble any physiological condition in the neuronal circuit, how is the NAergic system activated to modulate induction of homosynaptic plasticity and STDP under physiological conditions? Neuronal activity of LC neurons undergoes certain rhythms, which are closely related to the wake/sleep cycle of animals. They fire at highest frequency when animals are active waking, less when animals are quiet waking, and even less when animals are in non-rapid eye movement (REM) sleep and REM sleep [45, 46]. Thus, synaptic plasticity would be enhanced if an increase in synaptic activity in hippocampal circuit occurred simultaneously with an increase in the spiking rate of LC neurons; for example, synaptic plasticity might occur more easily during active waking. Sleep, however, has been suggested to play a significant role in declarative memory consolidation [47], as sleep deprivation has been shown to impair long-term synaptic plasticity significantly in the hippocampus in rats [48, 49]. These observations suggest that the temporal dissociation of NAergic activity from the neuronal circuit in the cortex and hippocampus during sleep could be essential for memory consolidation.

Like memory, LTP expression has different stages. There is an early-phase LTP (e-LTP), which usually lasts only 3–4 h after its induction and is independent of protein synthesis. Following e-LTP is late-phase LTP (l-LTP), which can last several hours in slice preparations or even days in vivo and is dependent on protein synthesis [50, 51]. At the cellular level, e-LTP might resemble the acquisition of new memory, and the conversion of e-LTP to l-LTP might resemble memory consolidation. As discussed above, the NAergic system, acting on β-adrenergic receptors, may have a permissive effect on the induction of e-LTP, and it may have a similar effect on memory acquisition. In line with this argument, a behavioral study showed that coeruleocortical NAergic lesions produced by intracerebral injection of 6-OHDA impairs learning behavior in animals [52]. What about the effect of the NAergic system on l-LTP? Does it also enhance l-LTP expression? Given that sleep is important for memory consolidation, which resembles the conversion of e-LTP to l-LTP at the cellular level, and that neuronal activity of LC neurons is low during sleep, it might appear that lower endogenous levels of NE would be the ideal situation for e-LTP to convert to l-LTP. However, experimental results do not support this argument. At mossy fiber synapses, homosynaptic e-LTP and l-LTP are both enhanced by β-adrenergic receptor activation [17]. One important factor that should be borne in mind is that mossy fiber LTP is NMDA receptor-independent [53, 54]; furthermore, the effect of endogenous levels of NE must be taken into account. Thus, at CA1 synapses, where LTP is NMDA receptor-dependent, the role of the NAergic system on l-LTP requires further investigation.

It has been shown that slow-wave sleep and sleep spindles are possible candidate mechanisms for sleep to enhance memory directly [55]. In hippocampal slices, it is also been demonstrated that somatic spiking at theta frequency [56] or ripple complex activity can directly enhance the conversion of e-LTP to l-LTP [57]. Thus, at the cellular and molecular levels, changes in the synchronization of the neuronal spiking pattern might be a crucial factor in enhancing expression of l-LTP and memory. Thalamic neurons can change their spiking patterns and degree of synchronized firing depending on the extracellular levels of NE, and the modification of the functions of ionic channels by NE can account for the changed spiking activity [52]. If the NAergic system did have a significant effect on l-LTP, would it involve a similar mechanism and allow somatic activity-dependent modification of l-LTP? Again, this would be an interesting question to answer.

Conclusions

A significant role of the NAergic system in modulating homosynaptic and associative forms of synaptic plasticity has been confirmed and the underlying molecular mechanism explored in detail. However, some questions remain to be answered, in particular the possible differential effects of the NAergic system on cortical plasticity during different states of brain functions, for example, during active awaking and/or sleep.

References

1. Bliss TVP, Collingridge GL (1993) Synaptic model of memory: long term potentiation in the hippocampus. Nature 361:31–39.
2. Malenka RC, Nicoll RA (1999) Long term potentiation: a decade of progress? Science 285:1870–1875.
3. Bear MF, Abraham WC (1996) Long-term depression in hippocampus. Annu Rev Neurosci 19:437–462.
4. Roberson ED, English JD, Sweatt JD (1996) Second messengers in LTP and LTD. In: Fazeli MS, Collingridge GL (eds) Cortical plasticity: LTP and LTD. BIOS Scientific, Oxford, pp 35–60.
5. Bienenstock L, Cooper NL, Munro PW (1982) Theory for the development of neuron selectivity: orientation specificity and binocular interaction in visual cortex. J Neurosci 2:32–48.
6. Bear MF, Kirkwood A (1996) Bidirectional plasticity of cortical synapses. In: Fazeli MS, Collingridge GL (eds) Cortical plasticity: LTP and LTD. BIOS Scientific, Oxford, pp 191–205.
7. Bear MF, Malenka RC (1994) Synaptic plasticity: LTP and LTD. Curr Opin Neurol 4:389–399.
8. Aston-Jones G, Cohen JD (2005) An integrative theory of locus coeruleus-norepinephrine function: adaptive gain and optimal performance. Annu Rev Neurosci 28:403–450.
9. Selden NR, Robbins TW, Everitt BJ (1990) Enhanced behavioral conditioning to context and impaired behavioral and neuroendocrine response to conditioned stimuli following ceruleo-cortical noradrenergic lesion: support for an attentional hypothesis of central noradrenergic function. J Neurosci 10:531–539.
10. Lin Y-W, Min M-Y, Chiu T-H, et al (2003) Enhancement of associative long-term potentiation by activation of β-adrenergic receptors at CA1 synapses in rat hippocampal slices. J Neurosci 23:4173–4181.
11. Kasamatsu T (1991) Adrenergic regulation of visual cortical plasticity; a role of locus coeruleus system. Prog Brain Res 88:599–616.
12. Yang H-W, Lin Y-W, Yen C-D, et al (2002) Change in bi-directional plasticity at CA1 synapses in hippocampal slices taken from 6-hydroxy-dopamine treated rats: the role of endogenous norepinephrine. Eur J Neurosci 16:1117–1128.
13. Richerson GB (2003) The autonomic nervous system. In: Boron WF, Boulpaep EL (eds) Medical physiology. Saunders, Philadelphia, pp 378–398.
14. Pertovaara A (2006) Noradrenergic pain modulation. Prog Neurobiol 80:53–83.
15. Katsuki H, Izumi Y, Zorumski CF (1997) Noradrenergic regulation of synaptic plasticity in the hippocampal CA1 region. J Neurophysiol 77:3013–3020.
16. Schimanski LA, Ali DW, Baker GB, et al (2007) Impaired hippocampal LTP in inbred mouse strains can be rescued by β-adrenergic receptor activation. Eur J Neurosci 25:1589–1598.
17. Huang Y-Y, Kandel ER (1996) Modulation of both the early and the late phase of mossy fiber LTP by activation of β-adrenergic receptors. Neuron 16:611–617.
18. Hopkins W, Johnston D (1988) Noradrenergic enhancement of long-term potentiation of mossy fiber synapses in the hippocampus. J Neurophysiol 59:667–678.
19. Bliss TVP, Goddard GV, Riives M (1983) Reduction of long-term potentiation in the dentate gyrus of the rat following selective depletion of monoamines. J Physiol (Lond) 334:475–491.
20. Loy R, Koziell DA, Lindsey D, et al (1980) Noradrenergic innervation of adult rat hippocampal formation. J Comp Neurol 189:699–710.
21. Moore RY, Bloom FE (1979) Central catecholamine neuron systems: anatomy and physiology of the norepinephrine and epinephrine systems. Annu Rev Neurosci 2:113–168.
22. Bronzino JD, Kehoe P, Mallinson K, et al (2001) Increased extracellular release of hippocampal NE is associated with tetanization of the medial perforant pathway in the freely moving adult male rat. Hippocampus 11:423–429.

23. Harley CW, Lalies MD, Nutt DJ (1996) Estimating the synaptic concentration of norepinephrine in dentate gyrus which produces beta receptor mediated long-lasting potentiation in vivo using microdialysis and intracerebroventricular norepinephrine. Brain Res 710:293–298.
24. Harik SI (1984) Locus coeruleus lesion by local 6-hydroxydopamine infusion causes marked and specific destruction of noradrenergic neurons, long term depletion of norepinephrine and enzymes that synthesize it, and enhanced dopaminergic mechanisms in the ipsilateral cerebral cortex. J Neurosci 4:699–707.
25. Lin Y-W, Yang H-W, Wang H-J, et al (2006) Spike-timing-dependent plasticity (STDP) at resting and conditioned lateral perforant path synapses on granule cells in the dentate gyrus: different roles of NMDA and group I metabotropic glutamate receptors. Eur J Neurosci 23:2362–2374.
26. Bi G, Poo M-M (2001) Synaptic modification by correlated activity: Hebb's postulate revised. Annu Rev Neurosci 24:139–166.
27. Dan Y, Poo M-M (2004) Spike-timing-dependent plasticity of neuronal circuits. Neuron 44:23–30.
28. Dan Y, Poo MM (2006) Spike timing-dependent plasticity: from synapse to perception. Physiol Rev 863:1033–1048.
29. Bea MF, Connors BW, Paradiso MA (2007) Neuroscience: exploring the brain (3rd ed). Lippincott Williams & Wilkins, Philadelphia.
30. Song S, Abbott LF (2001) Cortical development and remapping through spike timing-dependent plasticity. Neuron 32:339–350.
31. Song S, Miller KD, Abbott LF (2000) Competitive Hebbian learning through spike-timing-dependent plasticity. Nat Neurosci 3:919–926.
32. Debanne D, Gahwiler BH, Thomson SM (1998) Long term synaptic plasticity between pairs of individual CA3 pyramidal cells in rat hippocampal slice cultures. J Physiol (Lond) 507:237–247.
33. Feldman DE (2000) Timing-based LTP and LTD at vertical inputs to layer II / III pyramidal cells in rat barrel cortex. Neuron 27:45–56.
34. Chetaovich DM, Sweatt JD (1993) NMDA receptor activation increases cyclic AMP in area CA1 of the hippocampus via calcium/calmodulin stimulation of adenylyl cyclase. J Neurochem 61:1933–1942.
35. Makhinson M, Chotiner JK, Watson JB, et al (1999) Adenylyl cyclase activation modulates activity-dependent changes in synaptic strength and Ca^{2-}/calmodulin-dependent kinase II autophosphorylation. J Neurosci 19:2500–2510.
36. Otmakhova NA, Otmakhov N, Mortenson LH, et al (2000) Inhibition of cAMP pathway decreases early long-term potentiation at CA1 hippocampal synapses. J Neurosci 20:4446–4451.
37. Cooper DMF, Mons N, Karpen JW (1995) Adenylyl cyclase and the interaction between calcium and cAMP signalling. Nature 374:421–424.
38. Liauw J, Wu LJ, Zhuo M (2005) Calcium-stimulated adenylyl cyclases required for long-term potentiation in the anterior cingulate cortex. J Neurophysiol 94:878–882.
39. Wang H, Ferguson GD, Pineda VV, et al (2004) Overexpression of type-1 adenylyl cyclase in mouse forebrain enhances recognition memory and LTP. Nat Neurosci 7:635–642.
40. Wang H, Pineda VV, Chan GC, et al (2003) Type 8 adenylyl cyclase is targeted to excitatory synapses and required for mossy fiber long-term potentiation. J Neurosci 23:9710–9718.
41. Malenka RC, Nicoll RA (1997) Silent synapses speak up. Neuron 19:473–476.
42. Yuan LL, Adams JP, Swank M, et al (2002) Protein kinase modulation of dendritic K^- channels in hippocampus involves a mitogen-activated protein kinase pathway. J Neurosci 22:4860–4868.
43. Mulkey RM, Endo S, Shenolikar S, et al (1994) Involvement of a calcineurin/inhibitor-1 phosphatase cascade in hippocampal long-term depression. Nature 369:486–488.
44. Mulkey RM, Herron CE, Malenka RC (1993) An essential role for protein phosphatases in hippocampal long term depression. Science 261:1051–1055.

45. Aston-Jones G, Bloom FE (1981) Activity of norepinephrine-containing locus ceoruleus neuron in behaving rats anticipates fluctuation in sleep-waking cycle. J Neurosci 1:876–886.
46. Foote SL, Aston-Jones G, Bloom FE (1980) Impulse activity of locus coeruleus neurons in awake rats and monkeys is a function of sensory stimulation and arousal. Proc Natl Acad Sci U S A 77:3033–3037.
47. Ellenbogen JE, Payne JD, Stickgold R (2006) The role of sleep in declarative memory consolidation: passive, permissive, active or none? Curr Opin Neurobiol 16:716–722.
48. Ishikawa A, Kanayama Y, Matsumura H, et al (2006) Selective rapid eye movement sleep deprivation impairs the maintenance of long-term potentiation in the rat hippocampus. Eur J Neurosci 24:243–248.
49. Kopp C, Longordo F, Nicholson JR, et al (2006) Insufficient sleep reversibly alters bidirectional synaptic plasticity and NMDA receptor function. J Neurosci 26:12456–12465.
50. Frey U, Huang YY, Kandel ER (1993) Effects of cAMP stimulate a late stage of LTP in hippocampal CA1 neuron. Science 260:1661–1664.
51. Huang YY, Kandel ER (1994) Recruitment of long lasting and protein kinase A dependent long term potentiation in CA1 region of hippocampus requires repeated tetanization. Learn Mem 1:74–82.
52. Steriade M, McCarley RW (1990) Brainstem control of wakefulness and sleep. Plenum, New York.
53. Weisskopf MG, Castillo PE, Zalutsky RA, et al (1994) Mediation of hippocampal mossy fiber long-term potentiation by cyclic AMP. Science 265:1878–1882.
54. Weisskopf MG, Nicoll RA. (1995) Presynaptic changes during mossy fibre LTP revealed by NMDA receptor-mediated synaptic responses. Nature 376:256–259.
55. Gais S, Molle M, Helms K, et al (2002) Learning-dependent increases in sleep spindle density. J Neurosci **22:**6830–6834.
56. Dudek SM, Fields RD (2002) Somatic action potential are sufficient for late-phase LTP-related cell signaling. Proc Natl Acad Sci U S A 99:3962–3967.
57. Behrens CJ, van den Boom LP, de Hoz L, et al (2005) Induction of sharp wave-ripple complexes in vitro and reorganization of hippocampal networks. Nat Neurosci 8:1560–1567.

Section II
Stress and Fear

Involvement of the Amygdala in Two Different Forms of the Inhibitory Avoidance Task

Keng-Chen Liang[1,2], Chen-Tung Yen[2,3], Chun-Hui Chang[3], and Chun-Chun Chen[1]

Summary

The amygdala has been implicated in either mediating or modulating affective memory formation. To evaluate these two views, female Long-Evan rats received lesions of the amygdala central nucleus (CEA) or basolateral complex (BLC) and were trained on the step-through and step-down inhibitory avoidance tasks. Both responses learned under an inescapable procedure were impaired by CEA or BLC lesions, whereas those learned under an escapable procedure were impaired only by BLC lesions, suggesting that learning the two different tasks relied on the same amygdala circuit, but deploying different coping strategies engaged different circuits. To further address the issue, CEA and BLC ensemble activities were recorded in retention tests of an escapable step-down task and an inescapable step-through task. For both tasks, repeated exposure to a training apparatus caused habituation of neuronal firing, and training led to increased firing activities in both the CEA and BLC. Some neurons were highly reactive in one task, and others highly reactive in both. These data suggest that CEA and BLC neurons may be involved in processing both specific and general information inherent in inhibitory avoidance tasks and that the amygdala may play both modulating and mediating roles for different aspects of affective memory.

Key words Conditioning, Coping, Fear, Emotional memory, Single-unit activity

---

[1]Department of Psychology, National Taiwan University, 1 Roosevelt Road, Section 4, Taipei 10669, Taiwan

[2]Center for Neurobiology and Cognitive Science, National Taiwan University, 1 Roosevelt Road, Section 4, Taipei 10669, Taiwan

[3]Institute of Zoology, National Taiwan University, 1 Roosevelt Road, Section 4, Taipei 10669, Taiwan

Introduction

Evidence from both psychology and neuroscience suggests multiple systems of memory for humans and animals [1]. Compared to other subcategories, emotional memory, particular that of fear, receives much attention from scientists and lay people for its unique characteristics: It forms rapidly as attested by many one-trial affective learning tasks. Its durability sometimes lasts for the whole life-span of an organism. The memory is easily retrieved; and feeble cues elicit full-blown reexperiencing the fear. It serves a highly adaptive function by bestowing significance to the critical stimuli among those we encounter daily. By functioning normally, it keeps us away from danger. Yet in aberration, it is pathogenic and underlies various anxiety and affective disorders, such as the panic or posttraumatic stress disorder as well as depression.

When studying the neural bases subserving memory of emotional events in general or that of fear in particular, many researchers have conceived the amygdala as a sole site at which affective memory is formed and stored. Evidence conferring a role of forming or storing affective association to the amygdala can be summarized as follows: First, the amygdala receives convergent inputs of the conditioned and unconditioned stimuli; and it projects divergently to various subcortical nuclei controlling various fear-related responses [2]. Second, amygdala neurons show experience-driven neural plasticity, such as N-methyl-D-aspartate (NMDA) receptor-dependent long-term potentiation [3, 4]. Third, neuronal reactivity in the amygdala to a conditioned fear stimulus altered as learning progressed, and in some instances such changes modeled behavioral learning [5, 6]. Finally, manipulating the amygdala function shortly before or after training modulated acquisition or expression of the conditioned response [7, 8]. It should be noted that the above evidence is consistent with, but by no means proves, a role of the amygdala in storing fear association. What complicates this issue further is that the amygdala is implicated in processing various unconditioned emotional stimuli and the innate reactions to them [2], so a clear distinction between the effects of a treatment on learning and performance cannot be easily achieved. This constitutes a logical block for a clear interpretation of any effect induced by amygdala treatments.

An alternative view is that the amygdala serves a memory modulating function that is activated by emotion arousal and modifies memory formation processes elsewhere in the brain [9]. This view is suggested by the following lines of evidence: First, many treatments that perturb the amygdala function may either enhance or impair avoidance memory depending on several factors related to the treatment intensity and training conditions [10]. Second, these memory enhancing or impairing effects were attenuated by blocking the amygdala input–output pathway—the stria terminalis [11]—suggesting that treatments applied to the amygdala affect memory through its output influences on the connected areas. Third, whereas amygdala lesions applied shortly after training yielded robust amnesia, the same lesions applied weeks after training induced a negligible effect [12]. Further evidence revealed that pretest intra-amygdala infusion of lidocaine or CNQX impaired performance in a 1-day test but had no effect on a 21-day test [13, 14]. This temporally graded retrograde amnesia induced by posttraining lesions or pretest suppression of the amygdala suggests that memory for an inhibitory avoidance response could not be permanently stored in the amygdala. Fourth, learned responses eliminated by pretraining amygdala lesions could be readily reacquired and showed saving [15, 16], suggesting survival of some memory traces after the lesion. This view is inferentially more conservative and logically more defensible; nevertheless, it leaves open the kernel questions of where the neural trace for emotional memory might be located and how it could be identified.

It should be noted that the mediation view is mainly based on evidence accrued from Pavlovian fear conditioning, in which a specific stimulus is presented with an aversive event in an unique context, afterwards the organism shows fear responses such as increased freezing, heart rate, blood pressure, defecation, or urination in the presence of either the specific stimulus or the context. On the other hand, the modulation view gains much of its support from findings on inhibitory avoidance tasks. In one form of it, a rat is put into the lit side of a straight alley, and entry into the dark end due to a natural tendency of the rat is punished by an immediate and inescapable foot shock. Thereafter, the rat refrains from entering the dark side. The fear conditioning task and the inhibitory avoidance task share certain characteristics of aversively motivated learning: They are easy to train and test—both can be accomplished in a single trial; furthermore, memory forged as such is reliable and long-lasting on the one hand but malleable to modification of various treatments on the other. These similarities once led to a proposal that the inhibitory avoidance task essentially measures conditioned freezing, similar to any other Pavlovian fear conditioning task [17]. Yet our data did not support this view.

The inhibitory avoidance task differs from Pavlovian fear conditioning in several aspects: Fear conditioning is a pure classical conditioning task; and the degree of learning is expressed in forms of reflexive behavior under rigorous control dictated by the contingency between the conditioned and unconditioned stimuli. Although it may allow the unconditioned stimulus to be more adaptively processed [18], there is little room for subjects to change the contingency actively. Conversely, inhibitory avoidance learning contains components of both classical and operant conditioning and acquires a flexible response [14] that models human declarative memory [19]. The latter feature allows this task to assess, at both behavioral and neural levels, the effects of coping that is critical for determining the consequences of stress.

To distinguish between these two views requires evidence from both manipulative and correlative studies, and it would be helpful to obtain both types of data in two different tasks. The rationale is as follows: A site for an encoding and storing engram must show neural activity bearing both a causal and a correlative relation with the specific memory. For a site of modulation, however, its integrity during learning is critical for memory formation, but its neuronal activity in a retention test may or may not be correlated with performance. Furthermore, if the amygdala indeed stores specific information, recollecting two different memories is expected to activate two sets of amygdala neurons or one set of neurons with two response patterns. However, if the amygdala codes only general arousal instigated by different kinds of aversive stimuli and plays a modulating role, neurons in the amygdala should respond nonspecifically when recalling two different memories. Of course, these two views need not be mutually exclusive: The amygdala may not only mediate specific information coding in a task but also subserve a modulatory role in many tasks.

Involvement of Different Amygdala Nuclei in Two Inhibitory Avoidance Tasks: An Influence of Coping

Our laboratories have illustrated that inhibitory avoidance contains both classical and operant conditioning components [14] and recently further illustrated the influence of various training procedures on its memory at both behavioral and neural levels. In a conventional paradigm of the inhibitory avoidance task, a rat cannot escape from the shock, so active coping is impossible. However, in view of the importance of coping in determining the impact of a stressful experience, we made the shock escapable by leaving the apparatus door open during shock administration. Both inescapable and escapable training procedures involve operant conditioning as the rat receives punishment for its voluntary behavior. This inhibitory avoidance response can also be trained with a classical conditioning paradigm in which a rat is directly placed into the lit side of the apparatus; then the light goes off, and a shock comes on [14]. An electric circuit was devised to yoke three training alleys together for shock administration such that in each trio set of groups (the escapable, inescapable, and classical conditioning groups), rats received an equal amount of training foot shock.

The results show that in the retention test rats trained with the escapable and inescapable paradigms yielded similar degrees of avoidance behavior, but rats trained with the classical conditioning paradigm yielded a weak avoidance response. These data suggest that the same inhibitory avoidance response can be acquired via distinct learning processes, and the difference between operant and classical learning could be behaviorally dissociated. Yet a difference between the escapable and inescapable training procedure was not apparent unless the neural substrate was perturbed, as depicted in the following experiment.

We applied electrolytic lesions of the central (CEA), lateral, or basolateral amygdala nuclei to various groups of rats and trained them with the escapable and inescapable inhibitory avoidance tasks. Because lesions of the lateral or basolateral amygdala nuclei yielded almost identical results, they were pooled into a single group denoted as the basolateral complex lesions group (BLC). A 1-day retention test indicated that memory acquired under the escapable procedure was impaired by the BLC lesions but not by the CEA lesions. On the other hand, memory acquired with the inescapable paradigm was impaired by both BLC and CEA lesions. To test the generality of these findings, we re-examined the effect of such lesions on a step-down inhibitory avoidance task. In this task a rat was placed on a platform inside a chamber, and it would spontaneously step down to explore the floor. In a training trial, the rat stepped onto a hotplate preheated to 50°C as it descended. After this experience, the rat remained on the platform for a longer period of time before descending. The above groups of rats were subjected to this step-down inhibitory avoidance task in which during training the hotplate was made inescapable (by taking the platform away after stepping down) or escapable. The same results were replicated: the BLC lesions impaired memory acquired via the escapable and inescapable procedures of training, yet the CEA lesions impaired only memory acquired via an inescapable procedure.

These results showed that in two forms (step-through and step-down) of the inhibitory avoidance task, whereas formation of memory for the inescapable task required integrity of the CEA and BLC, that for the escapable task required the BLA but not the CEA. Thus, the manner of coping when encountering an aversive event has a critical role in pivoting the engaged neural circuitry in the amygdala. In view of the findings that an inescapable inhibitory avoidance task contains both classical and operant conditioning of fear [14] that respectively engaged the central and basolateral amygdala nuclei [20], the present data appeared to suggest that an opportunity of actively escaping from the shock during an inhibitory avoidance task enhanced the contribution of operant conditioning but diminished that of classical conditioning to the task.

Amygdala Neuronal Activity During Retention Test of Avoidance Responses

The above data showed that a rat was able to perform two different inhibitory responses motivated by fear engendered from two different aversive stimuli in two distinctive contexts. Furthermore, memories for escapable and inescapable experiences require the BLC, but that for the inescapable one also requires the CEA. These findings offer a ground to distinguish between the two views concerning the role of the amygdala in fear memory. A previous study showed that in an olfactory discrimination task subsets of neurons in the basolateral amygdala fire differentially to odors associated with a positive or a negative outcome, but no neuron fired to both types of odor [21]. Thus, appetitive and aversive responses could rely on two distinctive populations of amygdala neurons. However, no previous study to date has examined whether two different inhibitory avoidance responses acquired via different aversive stimuli in different contexts would also engage two populations of amygdala neurons.

To address this issue, we recorded ensemble unit activity in the amygdala of rats as they were tested for the two inhibitory avoidance tasks. Rats were trained on both the step-through and step-down inhibitory avoidance tasks motivated by either shock or heat. The procedure was as follows: From day 1 to day 3, rats went through 3 days of acclimation for the task apparatuses. They were first calmed down in a waiting cage for 5 min; neuronal activity at this pretask period was recorded as the baseline. Then they were allowed to explore the task apparatus freely for

5 min, this was designated the task period. After this period, they went back to the waiting cage and stayed for a posttask period of 5 min before returning to the home cage. Following the third acclimation session, rats were trained on the tasks sequentially in two consecutive days. To enhance distinction between the two tasks, the step-down task was trained with an escapable procedure, but the step-through task was trained with an inescapable procedure. A control group went through the same acclimation procedure with the experimental group in the same apparatus but received the foot shock and hotplate experience in different apparatuses located at another room. Retention performance of the two tasks for all rats was tested 1 day after training in the same apparatuses. For the test, the rat was placed in the lit side or on the platform and allowed 5 min of free moving in either apparatus.

Chronic electrodes made of eight microwires (50 μm) in a bundle type were implanted into the CEA or BLC region with stereotactic surgery. Unit activity was recorded in the pretask, task, and posttask periods of the first and third acclimation sessions and the test session. Electrophysiological data of a free-moving rat were recorded and processed by a Plexon system. Single units were screened out by a principal component analysis and template-matching algorithm, as reported in a previous article depicting the detailed methods and part of the data [22].

Behaviorally, rats in the experimental group readily acquired both types of avoidance behavior; they showed longer duration of staying in the lit side or on the platform after training than during the first or third acclimation period, whereas rats in the control group failed to do so (Fig. 1). This memory lasted at least for 10 days if rats were tested repeatedly. In both tasks, the averaged activity of all recorded units showed significant but small changes among the various behavioral sessions: Firing rates increased as a rat first entered the training apparatus in both experimental and control groups. The difference subsided during the third acclimation session for both groups. After training, rats in the experimental group showed increased activity during the task and posttask periods in comparison with the pretask period. Figure 2 shows the average ensemble activity recorded during the step-down test; a similar pattern has been reported for the step-through test [22].

The magnitude of overall firing increase did not bear any significant relation to the retention performance. The mean normalized increment over the pretask baseline of all neurons recorded in an individual animal was calculated for the task period during the test session of the step-through or step-down task. This increment score was not correlated with the time stayed on the platform in the step-down task or the time spent in the lit side of the shuttle box in the step-through task. Thus, the overall normalized firing increment of neurons recorded from individual animals, although statistically significant, could not predict the retention performance of that animal in either task. It is likely that the recorded neurons were functionally heterogeneous, and their discharges might be specifically linked to different behavioral parameters; thus little or no correlation could be detected by averaging firing over a long test period. This led us to seek for a correlation between the behavioral states and firing patterns of individual neurons.

Different categories of activity changes were detected in amygdala neurons in relation to behavioral changes. Taking what was found in the step-through task

Fig. 1. **A** Effect of foot shock training on the total time (seconds) stayed in the lit side of the step-through apparatus in the retention test. **B** Effect of hot plate training on the total time (seconds) stayed on the platform of the step-down apparatus in the retention test. Performance is expressed as the median and the interquartile range of each group. *con*, controls; *exp*, experimental animals. *$P < 0.05$

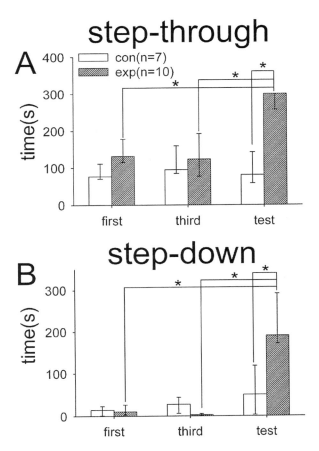

from our previous publication [22] as an example, habituation to novelty was one type of change: The unit fired vigorously during the rat's first encounter with the apparatus, with the discharge subsiding over the acclimation sessions as the rat explored the dark chamber. After foot-shock training, the rat rarely visited the dark side during the test; if it did, it stayed there only briefly. The unit showed no clear activity increase during the task period compared to the pretask period. Another type of unit showed learning-related activity changes. The rat explored the alley substantially, but the recorded unit showed little activation during the first to third acclimation periods. The rat no longer entered the lit side after the shock, but the unit increased its firing substantially in the test.

A third kind of unit showed both habituation and learning-related changes, as indicated in Fig. 3. The rat reduced its exploration in the dark side over the acclimation sessions and completely refrained from entering the darkness after training. The recorded unit showed vigorous firing as the rat first encountered the apparatus (Fig. 3A), but the activity gradually habituated over the acclimation sessions

Fig. 2. Average activities (mean ± SEM) of single units during the pretask (*pre-*), task, and posttask (*post-*) periods over the three task sessions for (**A**) the control (*con*) and (**B**) experimental (*exp*) groups in the step-down avoidance task. *$P < 0.05$

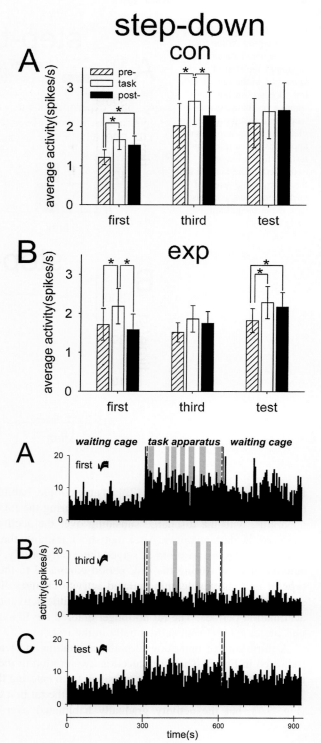

Fig. 3. A unit showing habituation in the discharge rate across the acclimation sessions and discharging vigorously after foot shock training in a retention test. The shaded vertical areas denote the time spent in the dark side

(Fig. 3B). However, it regained its vigorous firing mode in the retention test after foot-shock training (Fig. 3C). This type of activity change was consistent with previous findings that amygdala units altering their activity to the experience of learning were those initially responsive to the conditioned stimulus [23].

In many instances, amygdala units showed temporal modulation of firing rates within a test session. Because some experimental rats did shuttle back and forth between the lit and dark compartments, it would be interesting to know whether the activity level had any relationship with the rat's behavior in the test chamber. If the sustained ensemble activity recorded via a bundle of electrodes from the amygdala of a rat was scrutinized against its behavior in the test chamber, no clear relation was detected. To better delineate the relationship between transient firing changes and the behavioral state, we compared the 5-s ensemble firing rates just before and after a movement (either from the lit side to the dark side or vice versa) as the rat shuttled in the apparatus. An increase, no change, or decrease of the ensemble discharge rate was observed. An increase or decrease of the amygdala ensemble firing rate occurred more or less evenly when a control animal moved into or away from the dark side. In contrast, the amygdala ensemble activity was more likely to decrease as an experimental rat stepped into the dark chamber and to increase as it retreated. Thus, the lit side appeared to be associated with a higher overall firing rate in amygdala neurons and the dark side with a lower rate in the experimental group [22].

However, as we assessed the contingency between the activity of individual neurons in the amygdala and the behavioral state of experimental and control rats, we found three types of relation [22]: One type of units showed vigorous firing as a rat stayed in the lit side but decreased radically as it walked into the dark side. Another type of units showed little activity as a rat stayed in the lit side but fired vigorously as it entered the dark side. There were also units that modulated their firing pattern in a manner not related to the place in which the rat was located. Given the prominent firing preference in some amygdala neurons, one may wondered whether the preference was preexisting or acquired. The data showed that both exist. For some units, the firing preference was clear during the first acclimation phase, habituated to a negligible level during the third acclimation phase, and reappeared at the testing phase. For other units, there was no preference during the acclimation periods, but preference for the dark side appeared at the test phase. Thus, the firing preference in the amygdala neurons could exist either before or after inhibitory avoidance training. Of course, it would be difficult to determine whether the preexisting preference is innate or acquired before the assigned learning task and whether such preference bears a causal relation with the step-through behavior.

Previous findings have shown that acquisition, consolidation, or expression of an aversive memory in albino rats was more vulnerable to disruption of the right amygdala [24, 25]. However, we failed to find any differential responsiveness of neurons in the left or right amygdala during the test session in the experimental group. A possible interpretation for the inconsistency is that manipulation applied to the right amygdala may exert its effect by modulating neural activity elsewhere in the brain subserving memory storage rather than by altering in situ activity. Alternatively, the left and right amygdalae are intimately connected such that increased firing in one side would lead to an increase in the other, although only activity at one hemisphere bears a causal relation with memory. Whatever the reason may be, the data caution us against overinterpreting the correlation between neural activity changes and learning.

Correlated Activity of Amygdala Neurons in the Two Inhibitory Avoidance Tasks

In view of the above data, we went on to examine whether amygdala neurons respond similarly to recalling memories forged by both tasks. To assess the degree of similarity in neuronal responses under the two test conditions, we first calculated the correlation coefficients for activities of recorded neurons between the two tasks. For the control group, a high and significant correlation in the mean firing rates for individual neurons between the two tasks was detected during the task period in the first acclimation session, the third acclimation session, and the test session (r = 0.83, 0.91, and 0.81, respectively). For the experimental group, a significant correlation in the mean firing rates between the two tasks was detected during the first and third acclimation periods (r = 0.31 and 0.20, respectively), but it became negligible during the test session ($r = -0.01$). The data were understood as suggesting that before administration of an aversive stimulus neurons sampled from the amygdala of either the control or experimental rats reacted rather coherently in the two task environments. However, once different types of aversive stimuli were applied to each training apparatus in the experimental condition, the sampled units reacted differently, so the overall activity induced by the two tasks no longer correlated. Previous data [26] have shown that amygdala neurons in cats responded only to sensory stimuli that were emotionally significant; our findings support and extend this notion by showing that the amygdala neurons as a whole would not differentiate two sets of contextual or proprioceptive stimuli unless those stimuli acquired emotional values to signify different aversive events. These data might be understood to be congruent with the view that amygdala neurons code specific affective information.

This view predicts that the act of recalling during the two inhibitory avoidance tasks may activate different sets of amygdala neurons. However, when we looked into the activity of individual neurons in relation to the two tasks, three patterns of firing were observed. As shown in Fig. 4, some units did not respond in the step-down task but had increased activity in the step-through task (unit A) or vice versa (unit C), and some units showed increased activity during both tasks (unit B).

To address this question quantitatively, we selected highly responsive units from the recorded population as follows: During the test session, the mean activity of the 5-min pretask period in the waiting cage was calculated as the baseline. Unit activities during the task period collapsed into 5-s bins were normalized to the mean

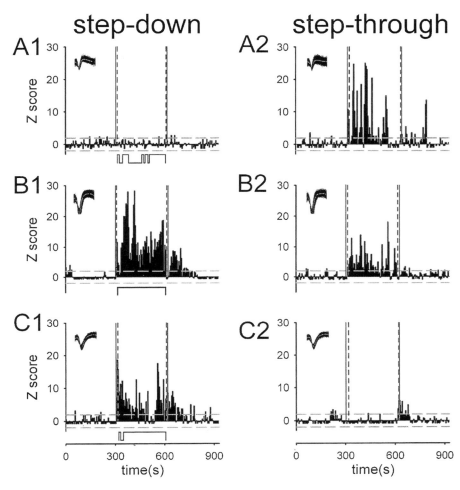

Fig. 4. Three example single units. Unit A was specifically responsive in the step-through avoidance task (**A**); unit B was responsive in both tasks (**B**); and unit C was specifically responsive in the step-down avoidance task (**C**). *Lines* underneath the task periods indicate the time at which the rat was on the small platform (*up position*) or on the hotplate (*down position*) during the step-down avoidance task. Rats stayed in the lit compartment all of the time in the step-through avoidance task. *Dashed horizontal lines* indicate ± 2.58 SD

baseline firing rate of the pretask period with the standard Z-score transformation. The mean of the Z-score averaged over the 60 bins during the task period was calculated for each unit, and those with a Z-score mean over 2.58 (3 standard deviations above the mean of baseline activity) were rated as highly responsive units. The distributions of the Z-score for units in the experimental and control groups are shown in Fig. 5. Only two units fit into this category in the control group: one activated in the step-through task and one activated in the step-down task.

Fig. 5. Distribution of unit responses during the test session for the two tasks in the control (**A**) and experimental (**B**) groups. The abscissas and ordinates are both Z-scores. *Dashed lines* in the horizontal and vertical directions indicate the 99% confidence levels. Response histograms of the three example units (**A, B, C**) are shown in Fig. 4

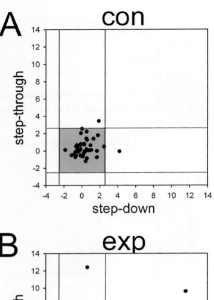

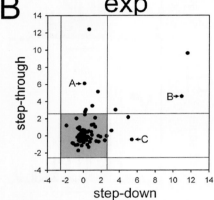

Conversely, in the experimental group, 12 units fit into the highly responsive category; among them, six were activated specifically in the step-through task, three were activated specifically in the step-down task, and three were activated in both tasks. This suggests that both specific and general responses were present. Although the number of neurons in each category may be too sparse to afford a firm conclusion, the data nonetheless suggest that in a retention test some amygdala units may respond to specific aspects of an individual task, but there are also amygdala neurons responding to both tasks. It would be interesting to know whether the neurons activated in both tasks showed similar or different firing patterns during the retention period. However, the scarce number of neurons recorded in each category at the present time prevented us from drawing any conclusions. Thus, neither the mediation view nor the modulation view on the role of the amygdala in aversive memory could be negated by these electrophysiological data recorded from the amygdala.

Some other analyses may also bear on the issue. In an initial attempt to augment the difference between the two tasks, the step-through task was trained with an inescapable procedure, whereas the step-down task an escapable one. Based on the data from our lesion study, one may wonder whether neurons in the CEA and BLC would show differential responsiveness in recalling these two avoidance responses. In the inescapable step-through task, neurons were equally activated in the CEA and BLC, which is consistent with findings that these two regions were both critical for the task. However, the activity difference between the two subregions was also absent for the escapable step-down task, which was unexpected from findings that only the basolateral amygdala complex was essential for this task. These data remind us once again that neuronal activity changes in specific amygdala nuclei may be a sufficient condition for locating the putative memory trace underlying the inhibitory avoidance response, but they may not be a necessary condition for normal operation of such a trace.

Another critical aspect relevant to the issue of whether the amygdala stores the trace of inhibitory avoidance memory is persistence of the neuronal changes over the retention period of memory. Previous data have shown that during the conditioning session of a fear response some neurons in the basolateral amygdala altered their activity during an initial phase but not during a late phase, whereas others altered their firing during a late phase but not during the initial phase [27]. Our preliminary data did show that some neurons were activated during the training session but not during the 1-day test session, which is consistent with the previous findings. Prolonged inhibitory avoidance memory in some animals allows us to evaluate the persistence of neuronal changes in the amygdala, which could not be addressed by the previous study. We have obtained electrophysiological data from two rats trained on the step-down inhibitory avoidance task and showed robust memory over a 10-day retention period. The recorded units showing increased overall discharges in a 1-day test failed to do so in a 10-day test. Negative results as such should be interpreted cautiously, as loss of the recorded neuron by the electrode or shift of the coded information to another set of neurons in the amygdala over the 10-day retention period could account for the lack of activity changes detected. However, they may suggest that after a long retention period the information related to an inhibitory avoidance task no longer relies on the amygdala. The latter view is consistent with our previous results that retention of an inhibitory avoidance response 12 days after initial acquisition was not impaired by pretest lesions of the amygdala [12] or pretest infusion of CNQX or lidocaine into the amygdala [13, 14]. Elucidation of this possibility should be further pursued in future research by recording sufficient numbers of neurons in the amygdala.

Conclusion

In an attempt to evaluate whether the amygdala plays a modulating or mediating role in the formation of emotional memory, a series of experiments were conducted to train rats on two inhibitory avoidance tasks; and the causal and correlative relations between the amygdala nuclei and retention of two avoidance responses were investigated. Results showed that rats trained with an escapable or inescapable procedure readily acquired the two inhibitory avoidance responses in two distinct environments. The neural circuitry engaged in the amygdala was sensitive to the execution of active coping behavior but impervious to the type of aversive stimuli or motor responses involved: In both the step-through and step-down tasks motivated by either electric shock or painful heat, acquisition and retention required the CEA and BLC during an inescapable training procedure but only the BLC during an escapable training procedure.

In contrast, both the CEA and BLC showed neuronal activity changes during either type of response. Several varieties of activity changes were noted: decreased firing during acclimation and/or increased firing after aversive learning. Activity of individual neurons appeared to bear more than one kind of relation with transient behavioral states, and some did react differentially in the two avoidance tasks whereas others reacted to both of them. Although the electrophysiological data collected in this study are consistent with previous findings that showed learning-driven neuronal plasticity in the amygdala [5, 6, 27], the profile of activity changes in the two tasks did not allow clear-cut discrimination between an interpretation of mediation and one of modulation. Furthermore, the nuclei and hemispheric distributions of activity changes do not match perfectly with results obtained from the manipulative experiments.

Evidence from both manipulative and correlative studies has shown that inhibitory avoidance learning engages brain structures beyond the amygdala [14]. A recent study showed that training on an inhibitory avoidance task induced hippocampal long-term potentiation that was responsible for subsequent avoidance memory [28]. Our laboratory has shown that the effect of altering amygdala functions on latent learning of an inhibitory avoidance response could be attenuated by suppressing the dorsal hippocampus [14]. Thus, even if plasticity in the amygdala has a causal link to inhibitory avoidance memory, it could not be an exclusive one. Evidence has shown that a simple conditioned defensive reflex may rely on a cascade of plastic changes developed successively in a series of brain structures from the sensory input to motor output [29–31]. If this is also the case for inhibitory avoidance learning, it would be critical to know how neural plasticity subserving the behavioral change initiates, builds up, and interacts in a distributed circuitry encompassing the amygdala. With this perspective, elucidation of whether and how the observed activity changes in amygdala neurons relate to changes in other brain regions also involved in formation of such memory may be more critical than proving or disproving whether they are the sole substrate of fear memory.

Acknowledgments The present study was supported by grants "Co-construction of Human Cognition, Neural Mechanism and Social Process" from the National Science Council of the ROC to K-C. Liang and C-T. Yen. The authors thank C-C. Huang, H-H. Chang, and other students for their participation in the behavioral experiments. Part of our data contained in this article have been presented elsewhere.

References

1. Schacter DL, Tulving E (1994) Memory systems of 1994. MIT Press, Cambridge.
2. Davis M (2000) The role of the amygdala in conditioned and unconditioned fear and anxiety. In: Aggleton J (ed) The amygdala: a functional analysis. Oxford University Press, New York, pp 213–287.
3. Gean PW, Chang FC, Huang CC, et al (1993) Long-term enhancement of EPSP and NMDA receptor mediated synaptic transmission in the amygdala. Brain Res Bull 31: 7–11.
4. Chien W-L, Liang K-C, Teng C-M, et al (2003) Enhancement of long-term potentiation by a potent nitric oxide-guanylyl cyclase activator 3-(5-hydroxymethyl-2-furyl)-1-benzyl-indazole. Mol Pharmacol 63:1322–1328.
5. Quirk GJ, Repa JC, LeDoux JE (1995) Fear conditioning enhances short-latency auditory responses of lateral amygdala neurons: parallel recording in the freely behaving rat. Neuron 15:1029–1039.
6. Quirk GJ, Armony JL, LeDoux JE (1997) Fear conditioning enhances different temporal components of toned-evoked spike trains in auditory cortex and lateral amygdala. Neuron 19:613–624.
7. Fanselow MS, Paulos AM (2005) The neuroscience of mammalian associative learning. Annu Rev Psychol 56:207–234.
8. Maren S (2001) Neurobiology of Pavlovian fear conditioning. Annu Rev Neurosci 24: 897–931.
9. McGaugh JL (2004) The amygdala modulates the consolidation of memories of emotionally arousing experiences. Annu Rev Neurosci 27:1–28.
10. Liang K-C, McGaugh JL, Yao H-Y (1990) Involvement of amygdala pathways in the influence of post-training intra-amygdala norepinephrine and peripheral epinephrine on memory storage. Brain Res 508:225–233.
11. Liang K-C, McGaugh JL (1983) Lesions of the stria terminalis attenuate the enhancing effect of posttraining epinephrine on retention of an inhibitory avoidance response. Behav Brain Res 9:49–58.
12. Liang K-C, McGaugh JL, Martinez JL Jr, et al (1982) Posttraining amygdaloid lesions impair retention of an inhibitory avoidance response. Behav Brain Res 4:237–249.
13. Liang K-C, Hu S-J, Chang SC (1996) Formation and retrieval of inhibitory avoidance memory: differential roles of glutamate receptors in the amygdala and medial prefrontal cortex. Chin J Physiol 39:155–166.
14. Liang K-C (2006) Neural circuitry involved in avoidance learning and memory: the amygdala and beyond. In: Jing Q, Rosenzweig MR, d'Ydewalle G, et al (eds) Progress in psychological science around the world. Vol 1. Neural, cognitive, and developmental issues. Psychology Press, Hove, UK, pp 315–332.
15. Kim M, Davis M (1993) Electrolytic lesions of the amygdala block acquisition and expression of fear-potentiated startle even with extensive training but do not prevent reacquisition. Behav Neurosci 107:580–595.
16. Maren S (1999) Neurotoxic basolateral amygdala lesions impair learning and memory but not the performance of conditional fear in rats. J Neurosci 19:8696–8703.
17. Cammarota M, Bevilaqua LRM, Kerr D, et al (2003) Inhibition of mRNA and protein synthesis in the CA1 region of the dorsal hippocampus blocks re-installment of an extinguished conditioned fear response. J Neurosci 23:737–741.
18. Domjan M (2005) Pavlovian conditioning: a functional perspective. Annu Rev Psychol 56:179–206.
19. Eichenbaum H (1997) Declarative memory: insights from cognitive neurobiology. Annu Rev Psychol 48:547–572.
20. Killcross S, Robbins TW, Everitt BJ (1997) Different types of fear-conditioned behaviour mediated by separate nuclei within amygdala. Nature 388:377–380.

21. Schoenbaum G, Chiba AA, Gallagher M (1999) Neural encoding in orbitofrontal cortex and basolateral amygdala during olfactory discrimination learning. J Neurosci 19:1876–1884.
22. Chang C-H, Liang K-C, Yen C-T (2005) Inhibitory avoidance learning altered ensemble activity of amygdaloid neurons in rats. Eur J Neurosci 21:210–218.
23. Ben-Ari Y, Le Gal La Salle G (1974) Lateral amygdala unit activity. II. Habituating and non-habituation neurons. Electroencephalogr Clin Neurophysiol 37:463–472.
24. Coleman-Mesches K, McGaugh JL (1995) Differential involvement of the right and left amygdala in expression of memory for aversively motivated training. Brain Res 670: 75–81.
25. Coleman-Mesches K, McGaugh JL (1995) Differential effects of pretraining inactivation of the right or left amygdala on retention of inhibitory avoidance training. Behav Neurosci 109:642–647.
26. O'Keefe J, Bouman H (1969) Complex sensory properties of certain amygdala units in the freely moving cat. Exp Neurol 23:394–398.
27. Repa KC, Muller J, Apergis J et al (2001) Two different lateral amygdala cell populations contribute to the initiation and storage of memory. Nat Neurosci 4:724–731.
28. Whitlock JR, Heynen AJ, Shuler MG, et al (2006) Learning induces long-term potentiation in the hippocampus. Science 313:1093–1097.
29. Cohen DH (1985) Some organization principles of a vertebrate conditioning pathway: is memory a distributed property? In: Weinberger NM, McGaugh JL, Lynch G (eds) Memory systems of the brain. Guilford Press, New York, pp 27–48.
30. Weinberger NM (2004) Specific long-term memory traces in primary auditory cortex. Nat Neurosci Rev 5:279–290.
31. Woody CD (1984) The electrical excitability of nerve cells as an index of learned behavior. In: Alkon D, Farley J (eds) Primary neural substrates of learning and behavioral changes. Cambridge University Press, Cambridge, pp 101–127.

Bruxism and Stress Relief

Sadao Sato[1-3], Kenichi Sasaguri[1-3], Takero Ootsuka[1-3], Juri Saruta[1-3], Shinjiro Miyake[1-3], Mari Okamura[1,3], Chikatosi Sato[1-3], Norio Hori[2,4], Katsuhiko Kimoto[2,4], Keiichi Tsukinoki[5], Kazuko Watanabe[2,6,7], and Minoru Onozuka[2,7,8]

Summary

The masticatory organ, originally developed as a branchial system, has evolved over a long period of geological time through a stage in which it was predominantly a tool for expressing aggression into an organ for emotional management. In humans, the strong grinding and clenching function of the masticatory muscles, known as bruxism, plays a role in mitigating stress-induced psychosomatic disorders by down-regulating the limbic system, the autonomic nervous system, and the hypothalamic-pituitary-adrenal (HPA) axis. Experimental research results showed that bruxism-like activity (BLA) has beneficial effects on stress-induced reactions, such as increased expression of Fos, neuronal nitric oxide synthase (nNOS), dual phosphorylated extracellular signal-regulated kinase (pERK1/2), corticotropin-releasing factor (CRF), and free radicals in the paraventricular nucleus (PVN) of the hypothalamus. It has also been shown to cause alterations in the blood neutrophil/lymphocyte ratio, adrenocorticotropic hormone (ACTH) level, and stomach ulcer formation in animals studies and has increased amygdala neuronal activity and salivary chromogranin A level in human studies. These findings strongly suggested that parafunctional activity of the masticatory organ—aggressive BLA

[1]Department of Craniofacial Growth and Development Dentistry, Division of Orthodontics, Kanagawa Dental College, 82 Inaoka-cho, Yokosuka 238-8580, Japan

[2]Research Center of Brain and Oral Science, Kanagawa Dental College, Yokosuka, Japan

[3]Research Institute of Occlusion Medicine, Kanagawa Dental College, Yokosuka, Japan

[4]Department of Oral and Maxillofacial Rehabilitation, Kanagawa Dental College, Yokosuka, Japan

[5]Department of Maxillofacial Diagnostic Science, Kanagawa Dental College, Yokosuka, Japan

[6]Gifu University, School of Medicine, Gifu, Japan

[7]Research Center of Health Promotion Dentistry, Kanagawa Dental College, Yokosuka, Japan

[8]Department of Physiology and Neuroscience, Kanagawa Dental College, Yokosuka, Japan

behavior—has the ability to decrease stress-induced allostatic overload. The health of the masticatory organ depends critically on occlusion, which must be of sufficient quality to carry out its important role in managing stress successfully. Occlusion and the brain must function in harmony. For these reasons, we must integrate the study of occlusion into the broader scope of medical science; in so doing, we can meaningfully advance the state of the art of dental care and general health care.

Key words Bruxism, Stress, Masticatory organ, Brain, Allostasis

Introduction

Bruxism, an activity that involves clenching, grinding, and/or gnashing of the teeth, is one of several nonmasticatory parafunctions of the masticatory organ that almost everyone does to some extent while asleep [1]. Sleep bruxism is also classified as a parasomnia, a sleep disorder that is not an abnormality of the processes responsible for sleep and awake states (American Academy of Sleep Medicine). The experimental results, however, have been consistent and unambiguous: when the masticatory muscle activity during sleep of supposedly "nonbruxing" subjects has been monitored, researchers have found that most of them do indeed flex their closing masticatory muscles [2–5]. Sleep bruxism episodes are subclassified into phasic, tonic, or both (mixed) types according to the burst duration of the jaw-closing muscle [6].

Degenerative occlusal conditions, including occlusal trauma, temporomandibular joint dysfunction, abfracture, attrition, cracks, and tooth migration are well known dental conditions that are somehow related to biomechanical load exerted by powerful bruxism activity [7–11]. Each component can withstand a certain amount of stress without damage; but when loads exceed some threshold value, tissues begin to be injured. When a tooth is lost, its work must be taken over by the teeth that remain, increasing the burden of each of them. The problem underlying clinical dentistry is that we, as occlusion care professionals, do not yet understand the significance and physiology of sleep bruxism.

Researchers have made progress in understanding why people brux during sleep, but theories are as yet incomplete. However, what is central to the discussion here is that a large number of people brux while they sleep, and they do not cause meaningful damage to their teeth or gums, suggesting that bruxing is not *abnormal*. Why is it that we humans brux? A hypothesis that has been gaining ground among researchers in medicine and dentistry is that humans use the masticatory organ to attenuate stress.

Animal research has shown that using the masticatory organ to express aggression prevents stress-induced reactions such as increased adrenocorticotropic hormone (ACTH) in serum [12], increased nicotinamide adenine dinucleotide (NAD) in brain [13], and formation of stomach ulcers [14, 15]. It is time to recognize this fundamental connection between psychology and dental occlusion, and it is also time to juxtapose the role of stress relief and the role of the masticatory organ and explore how their fusion can contribute to improving health care. In this chapter, we discuss hypotheses regarding stress, the brain, and bruxism in an attempt to stimulate the development of concepts and ideas and to lay the framework for a new field of dentistry [16, 17].

Evolutionary Aspects of the Masticatory Organ and Bruxism

The human masticatory organ, originally derived from one of the components of the branchial system, has evolved over a long period of time into an organ for emotional management, having passed through stages in which the organ was used predominantly as a tool or weapon for expressing aggression. During the process of evolution, as species adapted from life in the sea to life on land, the original branchial visceral organ evolved to form the face, pharynx, and masticatory organ. Phylogenetic relations have been preserved, and thus the human orofacial system basically retains the topography and function of the progenitor species. Their origin is evident from the nerve supply.

As a derivative of the first branchial arch, the masticatory organ has functionally changed from its original autonomic pumping role to an organ for expressing emotion. The trigeminal nerve supplies signal pathways for both efferent and afferent signals. During these processes, the masticatory organ has been mainly used for expressing emotion, particularly aggression, and for instinctive purposes such as predation. The masticatory organ would probably be considered to be directly related to the limbic system. Modern humans retain this connection, and therefore this organ is used more to express aggression in the form of sleep bruxism than it is used to masticate.

It was hypothesized that jaw movement increases the blood flow to anterior temporal lobe structures during acute activation of the limbic fear circuits. Jaw movement may increase the blood flow to temporal lobe structures by pumping blood through the temporal bone emissary veins, thus conferring a possible survival advantage during activation of the limbic fear circuits in expectation of situations requiring the freeze, flight, fight, fright, or fear acute response [18–22]. Bruxism, grinding and clenching, may primarily be a manifestation of experiencing acute fear or chronic emotional distress, respectively. We posit that the alleles, which wire the fear circuits to the brain stem nuclei activating the bruxism behaviors, were highly conserved in the human lineage, as they remained highly adaptive for modern humans.

Effects of Bruxism-like Activity in Animal Studies

Many animal species grind their teeth as a component of their response to a threatening or stressful situation. During the process of evolution, animals have long used the masticatory organ as a tool for emotional outlet rather than just a simple tool for chewing food [23, 24]. It has been suggested that modern humans, in effect, are still using the masticatory organ to express aggression if they are overwhelmed psychologically. Much research has shown that psychic stress and occlusal disharmony are related to bruxism [25, 26]. From a psychosomatic point of view, unresolved psychic problems are transferred to the level of organs. Utilizing the masticatory organ as a stress outlet is an efficient, risk-free solution to the problem of stress management [27]. Several pieces of evidence using animal models were proposed in recent years as described below, in which an activity of the masticatory organ showed an attenuating effect of stress-induced neurophysiological events.

Effects of Bruxism-like Activity on Stress-Induced Tissue Reactions

It has been found that experimental rats grind their teeth spontaneously as a component of their response to a restrained stress condition, suggesting use of the masticatory organ to relieve stress. Thus, in our animal study using rats, the duration of spontaneous bruxism-like activity (BLA) was measured by a specially designed electromyography system during a restraint stress condition. Levels of catecholamine, cortisol, and corticosterone and the neutrophil/lymphocyte ratio in the blood were measured after the condition had terminated; in addition, areas of hemorrhage in the gastric mucous membrane and the weights of the spleen and thymus were measured [28, 29].

The results showed that the duration of the BLA had significant protective effects on the alteration of the neutrophil/lymphocyte ratio in blood, the changes brought on by stomach mucosal damage, and atrophy of the thymus and spleen (Fig. 1). The plasma ACTH level was increased 30 min after undergoing restraint stress, while active biting (BLA) attenuated the ACTH level (Fig. 2). The neutrophil/lymphocyte ratio was reversed only 240 min after restraint stress application, which might be caused by damage to the internal organs such as the stomach, thymus, and spleen. The stress-induced decreasing lymphocytes and increasing neutrophils were recovered by active BLA in a duration-dependent manner (Fig. 3). More stress-related changes were seen in the rats whose spontaneous BLA time was short due to stress. In the stress-with-short-BLA animals, catecholamine, cortisol, and glucose levels in the blood were increased compared with those in the long-BLA animals (Fig. 4). The hemorrhaged area of the stomach in the short-BLA animals was larger than that in the long-BLA animals. All these stress-induced alterations of physiological parameters were attenuated by spontaneous BLA in a duration-dependent manner. The results suggested that a reaction of hypothalamus-pituitary-adrenal (HPA) axis and the sympathetic nervous system to the stress was regulated by spontaneous BLA of the masticatory organ.

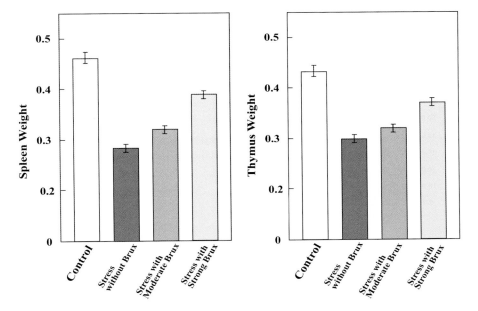

Fig. 1. Effects of restraint stress and bruxism-like activity (BLA) on the weight of lymphatic organs (spleen and thymus). Restraint stress was applied for 6 h. The stress-without-BLA animal showed a decrease in spleen and thymus weights, whereas this stress effect was attenuated in the stress-with-BLA animals. *Brux*, bruxism

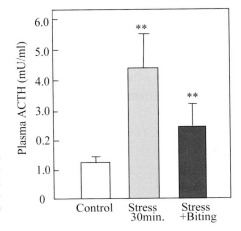

Fig. 2. Effects of restraint stress condition on the level of plasma adrenocorticotropic hormone (*ACTH*). Restraint stress increased plasma ACTH, whereas the BLA under stress condition prevented a stress-induced increase in the plasma ACTH level

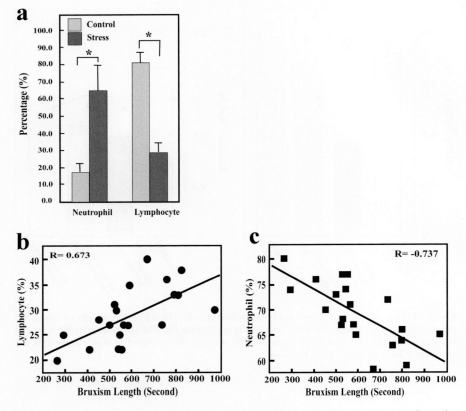

Fig. 3. Alterations of the neutrophil/lymphocyte balance in the blood leukocyte count. Restraint stress for 6h increased the number of neutrophils but decreased the lymphocytes (**a**). Alterations in the neutrophil/lymphocyte (**b/c**) ratio were correlated with the duration of BLA

Fos Expression in the Paraventricular Nucleus

Previous animal experiments have shown that biting significantly attenuates stress-induced increases in noradrenaline (norepinephrine) turnover in the hypothalamus, limbic areas, and amygdala [13]. Moreover, nonfunctional masticatory activity attenuates increases of striatal dopaminergic activity caused by tail pinch [30, 31]. It has been shown that psychological stress such as restraint or immobilization of the animals induces activation of c-*fos* and expression of Fos protein in the subsets of neurons in the central nervous system (CNS) [32]. It was of particular interest to us whether masticatory activity, such as biting a wooden stick (simulated BLA), might affect restraint-induced Fos protein expression in the rat brain stem.

Using immunohistochemical methods to detect Fos protein, we examined the effects of acute restraint stress and BLA on the neuronal circuits in the rat CNS

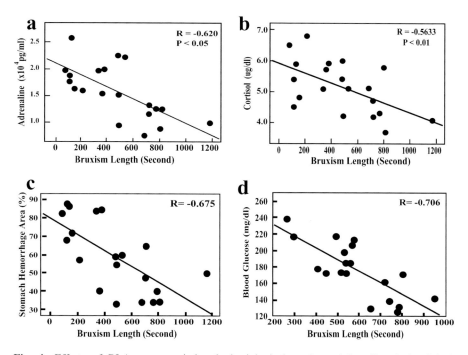

Fig. 4. Effects of BLA on stress-induced physiological markers. Adrenaline (epinephrine) (a) and cortisol (b) in blood were increased by the stress condition. BLA attenuated stress-induced changes in a duration-dependent manner. The area of gastric hemorrhage (c) and the blood glucose level (d) were increased by the stress condition, whereas BLA attenuated the stress-induced changes in a duration-dependent manner

[33]. Results showed that restraining rats in the supine position for 10–60 min invariably induced rapid expression of Fos protein in subsets of neurons in the somatosensory and motor cortices, paraventricular hypothalamic nucleus (PVN), locus coeruleus (LC), medial amygdaloid nucleus, and anterior portion of the paraventricular thalamic nucleus (Fig. 5).

In rats that were allowed to bite a wooden stick (BLA) during restraint, the enhanced Fos expression was significantly attenuated in all regions examined, with no further enhancement of the expression of Fos immunoreactivity in the facial, trigeminal, and vagal nuclei that innervate the skeletal muscles involved in the biting or nonmasticatory movement. The results corroborate that subsets of neurons in the rat brain are activated by acute restraint stress and express Fos protein. Suppression of Fos immunoreactivity by biting during restraint provides further support for the assumption that parafunctional masticatory activity attenuates adverse effects of psychological stress on the CNS. These results support the hypothesis that parafunctional masticatory activity attenuates the adverse effect of stress on neural circuits in the CNS.

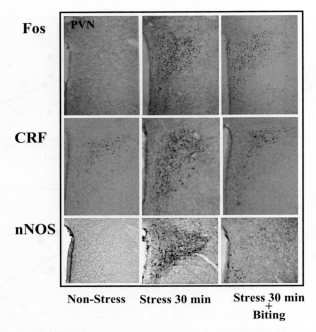

Fig. 5. Effect of biting (BLA) on expression of Fos, C-reactive protein (*CRF*), and nNOS in the paraventricular hypothalamic nucleus (*PVN*). Restraint-induced stress increased the number of positive nuclei. Active biting (BLA) suppressed the number of stress-induced positive nuclei

Corticotropin-Releasing Factor

Stress-induced activation of the HPA axis is characterized by enhanced expression of corticotropin-releasing factor (CRF) in the PVN and the consequent increases of ACTH and adrenal glucocorticoid in the plasma. Thus, the CRF expressed in the hypothalamus plays an important role in mediating behavioral responses to stressors. Restraining the body of an animal has been shown to activate and induce enhanced expression of CRF in the PVN of the rat hypothalamus.

We examined the effect of BLA on restraint-induced CRF expression in the rat hypothalamus [34]. The number of CRF-expressing neurons in the PVN increased significantly after a short time restraint (30 or 60 min) followed by a 180-min postrestraint period. Biting a wooden stick (BLA) during the restraint stress significantly suppressed the restraint-induced enhancement of CRF expression in the PVN (Fig. 5). Therefore, it is speculated that suppression of central noradrenergic transmission, including the afferent pathway mediated by LC, might be the mechanism by which CRF is suppressed by BLA. This antistress effect of the motor activity of the masticatory organ appears to be important, and this mechanism might be unconsciously in operation during an exposure to physical and psychological stressors, reducing the adverse effects of stress responses on the animal body.

Neuronal Nitric Oxide Synthase

Nitric oxide modulates the activity of the endocrine system in the behavioral response to stress. We investigated the effect of restraining the body of an animal on the expression of neuronal nitric oxide synthase (nNOS) in the PVN of the hypothalamus and the inhibitory effect of BLA on restraint-induced nNOS expression [35]. an increase in nNOS mRNA expression and nNOS-positive neurons in the rat hypothalamus was observed after 30 or 60 min of restraint stress. BLA during bodily restraint decreased nNOS mRNA expression in the hypothalamus. In addition, the number of nNOS-positive neurons was significantly reduced in the PVN of the hypothalamus (Fig. 5). These observations clearly suggest a possible anti-stress effect of masticatory activity (BLA) and how this mechanism might be unconsciously in operation during exposure to psychological stressors.

Phosphorylation of Extracellular Signal-Regulated Protein Kinase 1/2

As described above, acute immobilization stress induces Fos expression, which is generally used as a marker for neuronal activity. It has been suggested that it would be linked to phosphorylation of extracellular signal-regulated protein kinase 1/2 (pERK1/2) in the hypothalamic PVN. An immunohistochemical study was designed to determine whether acute immobilization stress induces pERK1/2 in the PVN and if the stress-induced pERK1/2 is attenuated by simultaneous BLA behavior [36]. Acute immobilization stress in up to 15-min increments produced detectable amounts of pERK1/2 that were proportional to the duration of the stress. BLA during the acute immobilization stress significantly reduced the amount of detectable pERK1/2 (Fig. 6). These results suggest that BLA activity during acute stress inhibits pERK1/2 in this region of the brain. It is feasible that the neuronal cellular response to acute stress is regulated, in some part, by pERK1/2 inhibition due to biting.

Free Radicals

Free radical reactions are involved in oxidative stress-mediated alterations in various physiological conditions. Because electron spin resonance (ESR) has been recognized as one of the most powerful techniques for detecting free radicals in biological tissues and cells, we have developed an ESR-based technique to detect free radical reactions in vivo. In vivo ESR spectroscopy and ESR imaging can be used to investigate redox status in living organisms noninvasively and in real time.

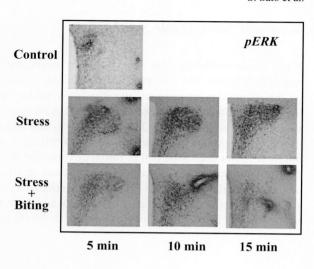

Fig. 6. Time course showing the difference of pERK1/2 induced in the PVN between two conditions—immobilization stress only and immobilization stress with concomitant biting (BLA). BLA decreased stress-induced pERK1/2 in the PVN

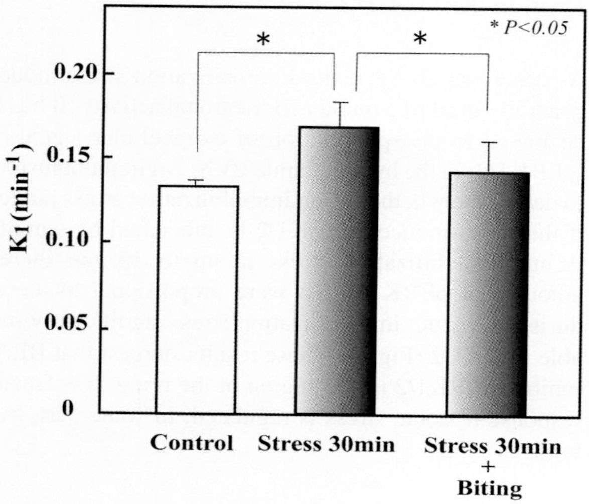

Fig. 7. Effect of biting (BLA) on free-radical production in rat brain. Restraint-induced stress increased the production of free radicals, whereas the stress-induced increase in free radical production was attenuated in the rats that were allowed to bite (BLA) during stress

We investigated the inhibitory effect of BLA on restraint stress-induced oxidative stress using these techniques [37]. A blood–brain barrier-permeable nitroxyl spin probe, 3-methoxycarbonyl-2,2,5,5,-tetramethylpyrrolidine-1-oxyl (MC-PROXYL), was administered to rats, after which L-band ESR and ESR-computed tomography (ESR-CT) imaging were used to show that the decay rate constant of MC-PROXYL in the hypothalamus of isolated brain after 30 min of restraint stress was more rapid than in unrestrained control rats (Fig. 7), which would suggest that restraint was associated with oxidative stress. Interestingly, biting (BLA) during restraint stress caused the decay rate constant of MC-PROXYL

Table 1. Summary of animal studies regarding with the effects of BLA on various stress-induced parameters

Parameter	Stress	Bruxism
Brain markers		
Fos	↑	↓
CRF	↑	↓
nNOS	↑	↓
pERK	↑	↓
Free radicals	↑	↓
Blood levels		
ACTH	↑	↓
IL1β	↑	↓
Corticosterone	↑	↓
Glucose	↑	↓
Leukocyte count		
Neutrophils	↑	↓
Lymphocytes	↓	↑
Thymus	↓	↑
Spleen	↓	↑
Stomach ulcer	↑	↓

BLA, bruxism-like activity; CRF, corticotropin-releasing factor; nNOS, neuronal nitric oxide synthase; pERK, phosphorylated extracellular signal-regulated kinase; ACTH, adrenocorticotropic hormone; IL1β, interleukin-1β

in isolated brain to approach that of the control group. These observations suggest that BLA suppresses oxidative stress induced by restraint stress, and that the anti-stress effect of masticatory motor activity movements, such as biting and chewing, are important for reducing the adverse effects associated with exposure to psychological stressors.

Table 1 illustrates the results from animal studies that have used rats as a stress model.

Human Studies

Functional Magnetic Resonance Imaging

The amygdala is reported to trigger physiological changes in response to a stressful situation [33]. If this is true, when the subject receives stressful information the amygdala should be activated. To test this hypothesis, we applied an auditory stimulus—a warning sound—to the subject throughout an earphone and measured blood oxygenation level-dependent (BOLD) signals using functional magnetic resonance imaging (fMRI). because a stressor is different for each individual, before we conducted the fMRI experiment we confirmed that the warning sound was, in fact, stressful. When the BOLD signals were statistically analyzed at

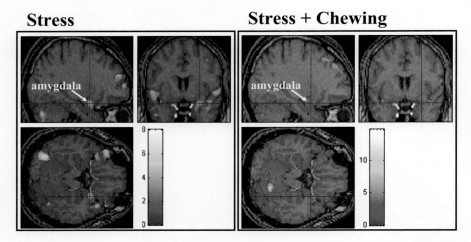

Fig. 8. Effect of chewing on the functional magnetic resonance imaging (fMRI) signal in the amygdaloid nucleus. The amygdaloid nucleus was activated in a subject stressed by being exposed to a warning sound. Allowing chewing (BLA) simultaneously with the activated amygdaloid nucleus by stress indicated that the masticatory activity attenuated stress-induced amygdaloid activation in humans

$P = 0.01$, the amygdala was markedly activated by this stimulus (Fig. 8, left). However, when the subject chewed gum (BLA) under the same stress condition, this sound-induced activation was mostly avoided (Fig. 8, right). Because the major efferents of the amygdala include projections to the septum and hypothalamus, providing not only a subcortical link for integrating limbic evaluative functions but also the switched HPA axis, chewing is associated with a reduced response to stress in the amygdala, thereby decreasing stress-stimulated hormonal activity.

Chromogranin A

Chromogranin A (CgA) is a 48-kDa acidic, hydrophilic protein that was originally isolated from chromaffin granules of the bovine adrenal medulla. later, CgA was found ubiquitously in the secretory vesicles of neuroendocrine cells. Previous studies have insisted on selective localization of CgA in sympathetic nerves, and radioimmunoassays that have revealed secretion of CgA in saliva have been used to evaluate stress conditions [38, 39]. Our previous study indicated that the strong immunoreactivity of CgA in human salivary glands was observed in the saliva matrix of ductal cavities and that CgA is produced predominantly by serous cells and secreted into saliva [40].

The present study sought to determine whether BLA influences the salivary CgA level. Samples of mixed saliva in a cotton roll from 44 adult healthy volunteers

Table 2. Effects of bruxism activity on the level of salivary chromogranin A in groups classified by their response to stress

Study group	CgA (pmol/mg protein), Exp/Cont		
	Resting	Stress	After 15 min
Group A (n = 16)			
Without brux	1.0	1.80 ± 0.70*	1.15 ± 0.51[§]
With brux	1.0	1.27 ± 0.71[††]	1.22 ± 0.49
Group B (n = 17)			
Without brux	1.0	1.57 ± 0.59*	2.22 ± 0.88*[§]
With brux	1.0	1.26 ± 0.45	1.23 ± 0.48[†]
Group C (n = 11)			
Without brux	1.0	0.58 ± 0.15*	0.60 ± 0.20*
With brux	1.0	1.10 ± 0.32	1.53 ± 0.57

CgA, chromogranin A; Exp/Cont, experimental / control
* Significant difference from Resting at $P < 0.01$ (Fisher's PLSD)
[§] Significant difference from Stress at $P < 0.01$ (Fisher's PLSD)
[†] Significant difference from without brux at $P < 0.01$ (paired t-test)
[††] Significant difference from without brux at $P < 0.05$ (paired t-test)

subjected (or not) to 5 min of stress were collected and analyzed for CgA content. Loud, unpleasant sounds (an emergency bell) were applied through headphones as a stressor. Salivary CgA was measured by an enzyme-linked immunosorbent assay (ELISA) using antibody against synthetic human CgA. Also investigated was the effect of BLA on the CgA level when stress was applied. Based on the response to the stressor, volunteers were divided into three groups: A, B, C. They showed different responses and different effects of BLA (Table 2).

This study demonstrated the possibility of evaluating stress levels of individuals by measuring salivary CgA. Salivary CgA responds to psychosomatic stress. Bruxism-like grinding activity prevented a stress-induced increase in salivary CgA in groups A and B, suggesting that the reaction of the parasympathetic–adrenomedullary system is regulated by the BLA of the masticatory organ.

Conclusion

Allostasis and the Masticatory Organ

The term "allostasis" has been coined to clarify ambiguities associated with the word "stress" [41]. Allostasis refers to the adaptive processes that maintain homeostasis through the production of mediators such as epinephrine, cortisol, and other chemical messengers. These mediators of the stress response promote adaptation during the aftermath of acute stress, but they also contribute to allostatic overload—the wear-and-tear on the body and brain that results from being stressed.

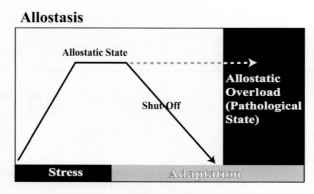

Fig. 9. Concept of allostasis. Allostasis refers to the adaptive processes that maintain homeostasis through changes. The stress response promotes adaptation in the aftermath of acute stress, but it also contributes to allostatic overload, the wear-and-tear on the body and brain that results from being stressed. This framework has helped demystify the biology of stress by emphasizing the protective as well as the damaging effects of the body's attempts to cope with the challenges known as stressors

This conceptual framework has created the need to know how to improve the efficiency of the adaptive response to stressors while minimizing overactivity of the same systems, as such overactivity results in many of the common diseases of modern life (Fig. 9). This framework has also helped demystify the biology of stress by emphasizing the protective as well as the damaging effects of the body's attempts to cope with the challenges known as stressors.

The terms "allostasis" and "allostatic overload" allow a more restricted, precise definition of the overused word "stress" and provide a view of how the essential protective and adaptive effects of physiological mediators that maintain homeostasis are involved in the cumulative effects of daily life when they are mismanaged or overused. These terms clarify inherent ambiguities in the concept of homeostasis. we also must note the ways in which they replace and clarify aspects of the "general adaptation syndrome" formulated by Selye [42]. Allostasis therefore distinguishes between the systems that are essential for life (homeostasis) and those that maintain these systems in balance (allostasis). When the systems involved in allostasis are elevated in a sustained manner, it is referred to as an allostatic state.

Experimental results described here suggest that masticatory activity plays a major role in effecting the restraint of stress-induced psychosomatic disorders by down-regulating the limbic system, the HPA axis, the autonomic nervous system (ANS), and the immune system. Therefore, occlusion of the masticatory organ contributes significantly to the individual's ability to manage allostasis. Bruxism in proper dentition can be recognized as a prophylactic system for all stress-related diseases (allostatic overload) (Fig. 10).

Allostasis and Masticatory Organ

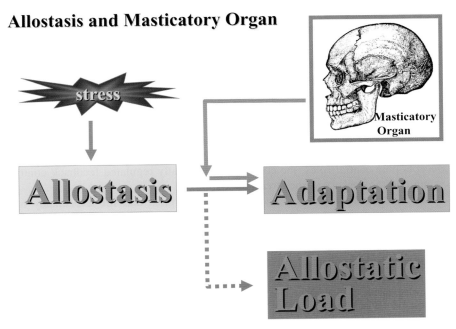

Fig. 10. Role of the masticatory organ in the allostatic response of the body. Masticatory activity plays a major role in effecting the restraint of stress-induced psychosomatic disorders by down-regulating the limbic system, the hypothalamic-pituitary-adrenocortical (HPA) axis, the autonomic nervous system (ANS), and the immune system. Therefore, the masticatory organ contributes significantly to the individual's ability to manage allostasis. Bruxism can be recognized as a prophylactic system for all stress-related diseases

Occlusion Medicine as a New Field of the Dentistry

The normal process of mastication consists of a series of highly coordinated related movements of the lower and upper jaws (and the mandibular and maxillary teeth they carry) along with coordinated actions of the lips, cheeks, tongue, temporomandibular joints, and other oral and perioral structures. Over and above this function of chewing and eating, the masticatory organ plays essential roles in speech, respiration, swallowing, maintaining posture, and stress management through the clenching and grinding of bruxism. Therefore, dentists must see more than just the static fit of upper and lower teeth as they diagnose and treat stomatognathic disorders, and they must recognize that they are custom-tailoring treatments to the complex physiological system of a unique individual.

The new field of dentistry, occlusion medicine, is that field of health care that accentuates the dynamic aspects of occlusion, accentuates the stress management function of the masticatory organ, and identifies and deals with stressors and stress-related diseases. It is an area that covers the whole being of the individual, including the psyche and the physical state of the total person. the challenge to the practitioner is one of changing a state of ill-health to that of well-being by applying well-chosen, effective dental treatment.

The human body is a set or arrangement of parts so related and connected as to form a unity, or an organic whole. it is considered a functioning organism with a number of bodily organs acting together to perform all of the functions necessary for life and health. In this context, the masticatory organ is closely related to the brain's limbic system and to expressing emotions as a functional unit. Therefore, we look forward to the day when dental practitioners view stress medicine and dental occlusion as two parts of an integrated whole—a field of study in which two concepts play central roles: the concept of a centrally regulated feedback control system and the concept of the stress–bruxism axis.

The masticatory organ has become more complex, and life expectancy has increased. The role of the masticatory organ as an emergency exit and stress valve must be recognized as the most important component of occlusion dentistry. The organ must prove its ability to execute its defensive responsibility; this is the center around which the concept of occlusion must be built.

Acknowledgments This work was performed at the Research Institute of Occlusion Medicine and Research Center of Brain and Oral Science, Kanagawa Dental College. It was supported by a grant-in-aid for Open Research from the Japanese Ministry of Education, Culture, Sports, Science, and Technology.

References

1. Arnold M (1981) Bruxism and the occlusion. Dent Clin North Am 25:395–407.
2. Sjoholm T, Lehtinen I, Helenius H (1995) Masseter muscle activity in diagnosed sleep bruxists compared with non-symptomatic controls. J Sleep Res 4:48–55.
3. Pierce CJ, Chrisman K, Bennett ME, et al (1995) Stress, anticipatory stress, and psychologic measures related to sleep bruxism. J Orofac Pain 9:51–56.
4. Kleinberg I (1994) Bruxism: aetiology, clinical signs and symptoms. Aust Prosthodont J 8:9–17.
5. Rugh JD (1991) Feasibility of a laboratory model of nocturnal bruxism. J Dent Res 70:554.
6. Lavigne GJ, Rompre PH, Montplaisir JY (1996) Sleep bruxism: validity of clinical research diagnostic criteria in a controlled polysomnographic study. J Dent Res 75:546–552.
7. Braem M, Lambrechts P, Vanherle G (1992) Stress-induced cervical lesion. J Prosthet Dent 67:718–722.
8. Coleman T, Grippo J, Kinderknecht K (2000) Cervical dentin hypersensitivity. Part II. Associations with abfractive lesions. Quintessence Int 31:466–465.
9. McCoy G (1999) Dental compression syndrome: a new look at an old disease. Oral Implantol 25:35–49.
10. Mandel L, Kaynar A (1994) Masseteric hypertrophy. NY State Dent J 60:44–47.
11. Spranger H (1995) Investigation into the genesis of angular lesions at the cervical region. Quintessence Int 26:149–154.
12. Weinberg J, Erskine M, Lavine S (1980) Shock-induced fighting attenuates the effects of prior shock experience in rats. Physiol Behav 25:9–16.

13. Tanaka T, Yoshida M, Yokoo H, et al (1998) Expression of aggression attenuates both stress-induced gastric ulcer formation and increases in noradrenaline release in the rat amygdala assessed by intracerebral microdialysis. Pharmacol Biochem Behav 59:27–31.
14. Guile MN, McCutcheon NB (1980) Prepared responses and gastric lesions in rats. Physiol Psychol 8:480–482.
15. Vincent GP, Pare WPD, Prenatt JE (1984) Aggression, body temperature, and stress ulcer. Physiol Behav 32:265–268.
16. Sato S, Slavicek R (2001) Bruxism as a stress management function of the masticatory organ. Bull Kanagawa Dent Coll 29:101–110.
17. Sato S, Yuyama N, Tamaki K, et al (2002) The masticatory organ, brain function, stress-release, and a proposal to add a new category to the taxonomy of the healing arts: occlusion medicine. Bull Kanagawa Dent Coll 30:117–126.
18. Yamamoto T, Hirayama A (2001) Effects of soft-diet feeding on synaptic density in the hippocampus and parietal cortex of senescence accelerated mice. Brain Res 902:255–263.
19. Onozuka M, Watanabe K, Mirbod SM, et al (1999) Reduced mastication stimulates impairment of spatial memory and degeneration of hippocampal neurons in aged SAMP8 mice. Brain Res 826:148–153.
20. Wilkinson L, Scholey A, Wesnes K (2002) Chewing gum selectively improves aspects of memory in healthy volunteers. Appetite 38:235–236.
21. Farella M, Bakke M, Michelotti A, et al (1999) Cardiovascular responses in humans to experimental chewing of gums of different consistencies. Arch Oral Biol 44:835–842.
22. Sasaguri K, Sato S, Hirano Y, et al (2004) Involvement of chewing in memory processes in humans: an approach using fMRI. In: Nakagawa M, Hirata K, Koga Y, et al (eds) Frontiers in human brain topography. International Congress Series 1270. Elsevier, Amsterdam, pp 111–116.
23. Every RG (1965) The teeth as weapons: their influence on behaviour. Lancet 10:685–688.
24. Every RG (1960) The significance of extreme mandibular movements. Lancet 2:37–39.
25. Labezoo F, Naeije M (2001) Bruxism is mainly regulated centrally, not peripherally. J Oral Rehabil 28:1085–1091.
26. Slavicek R (1992) Das sogenannte kauorgan als kybernetischer regelkreis — gesamtheit—liches verstandnis in der funktionslehre. Phillipine J 9:385–391.
27. Slavicek R, Sato S (2004) Bruxism—a function of the masticatory organ to cope with stress. Wien Med Wochenschr 154:584–589.
28. Takashina H, Itoh Y, Iwamiya M, et al (2005) Stress-induced bruxism modulates stress-induced systemic tissue damages in rat. Kanagawa Shigaku 40:1–11 (Japanese with English abstract).
29. Ishii H, Tsukinoki K, Sasaguri K (2006, in press) Role of the masticatory organ in maintaining allostasis. Kanagawa Shigaku (Japanese with English abstract).
30. Gomez FM, Giralt MT, Sainz B, et al (1999) Possible attenuation of stress-induced increases in striatal dopamine metabolism by the expression of non-functional masticatory activity in the rat. Eur J Oral Sci 107:461–467.
31. Areso MP, Giralt, MT, Sainz B, et al (1999) Occlusal disharmonies modulate central catecholaminergic activity in the rat. J Dent Res 78:1204–1213.
32. Chowdhury GM, Fujioka T, Nakamura S (2000) Induction and adaptation of Fos expression in the rat brain by two types of acute restraint stress. Brain Res Bull 52:171–182.
33. Kaneko M, Hori N, Yuyama N, et al (2004) Biting suppresses Fos expression in various regions of the rat brain: further evidence that the masticatory organ functions to manage stress. Stomatologie 101.7:151–156.
34. Hori N, Yuyama N, Tamura K (2004) Biting suppresses stress-induced expression of corticotrophin-releasing factor (CRF) in the rat hypothalamus. J Dent Res 83:124–128.
35. Hori N, Lee MC, Sasaguri K, et al (2005) Suppression of stress-induced nNOS expression in the rat hypothalamus by biting. J Dent Res 84:624–628.

36. Sasaguri K, Kikuchi M, Hori N, et al (2005) Suppression of stress immobilization-induced phosphorylation of ERK 1/2 by biting in the rat hypothalamic paraventricular nucleus. Neurosci Lett 383:160–164.
37. Miyake S, Sasaguri K, Hori N, et al (2005) Biting reduces acute stress-induced oxidative stress in the rat hypothalamus. Redox Rep 10:19–24.
38. Kanno T, Asada N, Yanase H, et al (1999) Salivary secretion of highly concentrated chromogranin A in response to noradrenaline and acetylcholine in isolated and perfused rat submandibular glands. Exp Physiol 84(6):1073–1083.
39. Kanno T, Asada N, Yanase H, et al (2000) Salivary secretion of chromogranin A control by autonomic nervous system. Adv Exp Med Biol 482:143–151.
40. Saruta J, Tsukinoki K, Sasaguri K, et al (2004) Expression and localization of chromogranin A gene and protein in human submandibular. Cell Tissues Organs 180:237–244.
41. McEwen B (2002) Sex, stress and the hippocampus: allostasis, allostatic load and the aging process. Neurobiol Aging 23:921–939.
42. Selye H (1936) Syndrome produced by diverse nocuous agents. Nature 138:32.

Section III
Pain

Muscular Pain Mechanisms: Brief Review with Special Consideration of Delayed-Onset Muscle Soreness

Kazue Mizumura

Summary

Muscular pain is quite common, but its mechanisms remain poorly understood. A number of experimental muscular pain models are reviewed herein, with emphasis on delayed-onset muscle soreness (DOMS). DOMS was selected for more detailed review because factors other than inflammation seem to be involved in clinically common muscle pain, and DOMS after lengthening contraction (LC) seems to involve such aspects. The discussion on the methodology of measuring muscle pain with transcutaneous pressure stimulation in humans and awake animals is reviewed; and then studies that have shown the existence of mechanical hyperalgesia in DOMS model animals produced by LC and elongation of the hyperalgesic period (for 5 days) in aged animals (130 weeks old) are introduced. In addition, the response characteristics of thin-fiber receptors and their roles in cardiorespiratory control during exercise and pain are reviewed. In addition, increased sensitivity of thin-fiber receptors to mechanical stimulation in hyperalgesic muscle after LC is discussed. Finally, a future direction for research using the DOMS model is proposed.

Key words Delayed-onset muscle soreness, Mechanical hyperalgesia, Nociceptors, Pain, Exercise

Introduction

Muscular pain such as tender neck, low back pain, or knee pain is quite common, and its incidence increases with age. Muscle pain is often related to movement and it is felt when the muscle is pressed (tenderness); that is, mechanical hyperalgesia is characteristic of muscle pain. Also typical of muscle pain is that painful

---

Department of Neuroscience II, Research Institute of Environmental Medicine, Nagoya University, Furo-cho, Chikusa-ku, Nagoya 464-8601, Japan

muscle often contains a stiff part or taut band, and pain often spreads to distant areas when a certain point (trigger point) in the taut band is compressed. Much less research effort has been directed toward the mechanisms of muscular pain compared with those of cutaneous pain, perhaps because induction and evaluation of muscle pain and/or mechanical hyperalgesia in awake animals/humans cannot be done without contaminating sensations from intervening skin and subcutaneous tissues.

Muscle Pain Models and Their Contribution to Understanding the Muscle Pain Mechanism

Animal models are essential for the study of neurophysiological mechanisms of muscle pain, and various models have been developed, including acute pain models produced by injection of capsaicin, hypertonic saline, and inflammatory mediators or by ischemia. An acute inflammation model has also been made by injecting carrageenin. Delayed-onset muscle soreness (DOMS) is a model with a little longer duration, but no complete models of chronic muscle pain have yet been developed.

Acute Muscle Pain Model

Intramuscular injection of chemicals is a better method for inducing muscle pain than mechanical compression of the muscle, in the sense that it can induce muscle pain without contaminating cutaneous sensations. This method has been used to determine if a certain substance induces muscular pain, which substances cause intense pain, the quality of the pain, where pain refers, and so on. Injection of hypertonic saline into the muscle was first used by Kellgren [1] to show the spread of pain (referral) to areas different from the site of injection. Later, it was used by a Denmark group (Arendt-Nielsen's group) to study the intensity and time course of pain as well as referred pain [2–5] and the corresponding changes in pathological conditions [6–8]. Hypertonic saline injected into the muscle can vigorously excite not only nociceptors but also other forms of thin-fiber receptors, which means that all types of muscle thin-fiber afferents can be excited by hypertonic saline [9, 10]. The mixed excitation of various sensory receptors might modify the pain sensation (not identified yet), but this method excludes concomitant excitation of cutaneous afferents.

Substances that appear during muscle contraction and inflammation, such as adenosine triphosphate (ATP) [11], adenosine [4], bradykinin [5, 12], glutamate [13, 14], and serotonin [5], have been injected into muscle, which has revealed that muscle afferents are sensitive to these substances and that they induce pain. Capsaicin is not an endogenous algesic substance, but it stimulates heat-sensitive nonselective cation channel TRPV1 [15] and causes pain when injected into the skin or viscera. The same is true for muscle [16]. The hyperalgesic and allodynic effects of some of the above substances [12, 14] and nerve growth factor (NGF) [17] have also been studied. In addition to chemicals, injection of heated normal saline has been used to reveal that muscle afferents are also sensitive to heat [18], and that it results in pain [19].

Ischemic Pain Model

Ischemia in muscle is well known to cause pain, a typical example of which is chest pain induced by cardiac ischemia. A model of cardiac ischemia has been made by constricting the coronary artery [20]. There is evidence that cardiac sympathetic A-δ and C-fibers are excited during myocardial ischemia [21]. Possible causes of pain that appear/increase during ischemia and sensitize nociceptors are adenosine, bradykinin, lactic acid, and low pH. Pan and Longhurst [22] recorded sympathetic afferents that were excited by ischemia but failed to show activation by adenosine. Bradykinin excited cardiac sympathetic C-fibers [23, 24], but ischemic cardiac pain was not suppressed by acetylsalicylic acid [24] even though bradykinin-evoked excitation was suppressed. Acid is now considered to be a more likely candidate for excitation of cardiac afferents to cause cardiac pain [20, 25].

There are two acid-sensing ion channels: TRPV1 and ASIC3. ASIC3 is more strongly expressed in cardiac afferents [25, 26] and is thus thought to play a major role in ischemia/acid-induced cardiac pain. Lactic acid is more potent than other forms of acid in inducing afferent excitation [20]. Supporting the hypothesis that ASICs mediate ischemic cardiac pain, ASIC3 current induced by acidic pH has been reported to be increased by lactic acid and ATP [27, 28], which also increase in the ischemic heart. It is necessary to examine whether the same is true in skeletal muscle.

Inflammation Model

Similar to other tissues, inflammation in the muscle causes pain. Injection of carrageenin [29, 30] and complete Freund adjuvant [31] have been used to induce inflammation. Changes in thin-fiber receptor activities in inflamed muscle are reviewed later in the chapter.

Inflammatory signs do not always exist in clinical cases of muscle pain/hyperalgesia, however. Therefore, models other than inflammation are needed. A neuropathic condition is thought to be one of the causes of muscle pain, but until now it has not been systematically examined in neuropathic pain patients or animal models of neuropathic pain produced by nerve injury.

Delayed-Onset Muscle Soreness Model

DOMS is described as an unpleasant sensation or pain after unaccustomed strenuous exercise. It is known that lengthening contraction (LC), also termed eccentric contraction, can induce DOMS more easily than shortening contraction (also termed concentric contraction) [32, 33]. In humans, DOMS is not apparent shortly after exercise but becomes apparent ~24h afterward and reaches a peak 24–48h later. It disappears within 3–7 days [33–35]. There is usually no spontaneous pain [35].

Histological [36–38] and ultrastructural [39, 40] studies on humans and animals have shown microinjuries in exercised muscle, and biochemical studies have found leakage of enzymes such as creatinine kinase (CK) and lactic dehydrogenase (LDH) from exercised muscle [36, 41, 42]. Invasion of inflammatory cells (macrophages) has not always been found in exercised muscle when the muscle was sore [43, 44]. One comprehensive human study did not detect any significant increase in inflammatory cytokines or macrophages up to 7 days after LC [45].

DOMS is usually subclinical, but it interferes with an athlete's performance, especially at the beginning of a new season [46]. It also has potential to lead to chronic injury and pain. On the other hand, DOMS might be another aspect of the adaptation process that leads to muscle hypertrophy [47]. Thus, causal agents for muscle mechanical hyperalgesia might be by-products on the way to muscle adaptation.

Brief Review of Hypotheses for the Mechanism of DOMS

Several theories have been proposed to explain the mechanism of DOMS. It should be mentioned here that the question of whether animals that undergo LC develop muscle mechanical hyperalgesia has not been examined until a recent study [48] and our 2005 report [49], and thus the pain mechanism of DOMS remains largely unknown.

Lactic Acid Theory

Lactic acid increases in exercised muscle, and its increase in plasma shortly after exercise has been reported [44]. However, the supposition that lactic acid by itself causes DOMS is strongly refuted by the fact that the higher degree of metabolism associated with concentric exercise has failed to result in delayed soreness [32, 50]. In addition, lactic acid levels return to preexercise levels within 1h following exercise. Therefore, lactic acid may contribute to the acute pain associated with fatigue following intense exercise, but DOMS cannot be attributed to it.

Spasm Theory

The spasm theory was proposed based on the observation that resting muscle activity [electromyographic (EMG) activity] was increased after eccentric exercise [51]. An observation apparently supportive of this theory is that after lengthening exercise of the upper arm the elbow joint angle becomes smaller (i.e., flexed position) [52], although EMG activity was not detected. This flexion was thought to lead to compression of local blood vessels, ischemia, and accumulation of pain substances, which initiated a "vicious cycle." However, reports on the presence or absence of EMG activity in the exercised muscle are controversial, as already described [51, 52].

Connective Tissue Damage Theory

It has been proposed that the connective tissue that forms sheaths around bundles of muscle fibers plays a role in DOMS. The content and composition of connective tissue differs among muscle fiber types; that is, type I (slow twitch) fibers display a more robust structure than type II (fast twitch) fibers. Fast twitch fibers may demonstrate increased susceptibility to stretch-induced injury, and excessive strain of the connective tissue may lead to muscle soreness.

Muscle Damage Theory

The muscle damage theory was proposed based on the finding that the contractile component of muscle tissue, particularly at the level of the Z-line, was disrupted following eccentric exercise. Characteristic microscopic/electron microscopic findings are Z-band broadening/streaming and there is focal disruption of the striated band pattern and disorganized sarcomeres [53, 54]. This damage is the result of increased tension per unit active area because fewer motor units are activated during eccentric (lengthening) exercise than during concentric (shortening) exercise to produce a given muscle force, and this increased tension could cause mechanical disruption of structural elements in the muscle fiber [33]. Demonstration of leakage of muscle enzymes (e.g., CK, LDH) supports this hypothesis [36]. It is interesting that two peaks in the enzyme leakage—one shortly after exercise and another about 40 h after exercise—were observed [36]. The time course of the second peak is compatible with the peak soreness (2 days after exercise in our model [49], described later), but the first is too early. Thus, muscle damage is not the direct cause of soreness, although the first peak seems to be related to the initiation of the entire process that leads to soreness. The meaning of the second peak is not known.

Inflammation Theory

The inflammation theory was proposed based on observations of edema and infiltration of inflammatory cells. The rapid breakdown of damaged muscle fibers and connective tissue, in addition to the accumulation of bradykinin, histamine, and prostaglandins, attracts monocytes and neutrophils to the injury site. However, the time course of inflammatory cell infiltration does not coincide well with that of soreness [43]. Some studies found no neutrophil accumulation [55]. An intensive, comprehensive study [45] on humans reported no difference in inflammatory markers between subjects who underwent downhill or uphill running and the controls (no exercise). Involvement of inflammation was also examined with the use of antiinflammatory drugs. However, the effects of antiinflammatory drugs differed among researchers and depended on whether a drug was administered prophylactically (before exercise) or therapeutically (after exercise) [46, 56].

It must be noted that most of the drugs used, even in reports with successful results, were cyclooxygenase (COX) inhibitors. Thus, effective suppression of DOMS by such drugs may not mean that an inflammatory process is involved in DOMS—only that production of prostaglandins is suppressed. It should be mentioned that prostaglandin synthesis in skeletal muscle can be induced by muscle stretching alone [57]. Moreover, eccentric exercise itself induces release of ATP and bradykinin, and eccentric exercise alone as well as bradykinin and ATP can activate mitogen-activated protein (MAP) kinases in muscle [58, 59]. MAP kinases are known to induce changes in expression of many substances, including COX, in dorsal root ganglion neurons [60], so it is plausible that the same may occur in muscle.

To date, no study has directly addressed mechanical hyperalgesia after LC in an animal model of DOMS. To address the neural mechanisms of DOMS, it is essential that we clarify the existence of mechanical hyperalgesia (tenderness, movement-related pain) in animals after LC. Reports have appeared, including ours, that demonstrate the existence of mechanical hyperalgesia in rats [49, 61] and rabbits [48], as well as increased mechanical sensitivity of thin-fiber afferents in hyperalgesic muscle 2 days after LC [10].

The following section touches on the methodology of detecting mechanical hyperalgesia in animals and the function of muscular afferents. Relevant studies are then introduced.

Existence of Tenderness in the Muscle of Rats that Underwent Lengthening Contraction

Methodological Consideration of Measuring Muscle Mechanical Withdrawal (Pain) Threshold

Transcutaneous pressure is usually used to measure muscle pain. Because this stimulus also affects the skin, however, it is difficult to attribute the origin of the evoked pain to muscle, especially in animals. The use of a larger probe, such as a Fischer algesiometer with a 10mm tip diameter has been recommended to measure muscle pain [62], but the validity or limitation of this method is not known. Moreover, it is not known which diameter should be used for animals. Therefore, we examined whether the pain threshold measured with different-sized probes was influenced by surface anesthesia.

Figure 1a shows the results obtained in humans [63]: The pain threshold measured with a probe having a tip diameter of 1mm was increased after surface anesthesia (Xylocaine patch; Wyeth-Lederle Japan, Tokyo, Japan) for 1h, which indicates that this probe measured the cutaneous mechanical pain threshold. A

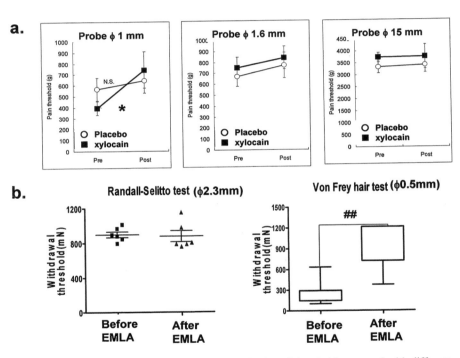

Fig. 1. Effects of surface anesthesia on the pain/withdrawal threshold measured with different-sized probes. **a** Results of the human experiment. Pressure pain threshold was measured with three types of pressure probe (diameters 1.0, 1.6, and 15mm) that were attached to an electronic pressure algometer and pressed to the skin surface over the brachioradial muscle. Surface anesthesia was performed by applying a piece of anesthetic adhesive containing Xylocaine. The pressure pain threshold measured with probes 1.6 and 15mm in diameter was not influenced by surface anesthesia, suggesting that these probes measure the deep pain threshold. (From Takahashi et al. [63], with permission) **b** Results of the rat experiment. A pressure probe was applied to the anterior surface of the lower hind leg. Surface anesthesia was obtained by applying EMLA (lidocaine/prilocaine) cream to the skin surface for 30min. The withdrawal threshold measured with the Randall-Selitto apparatus (tip diameter 2.3mm) was not changed by surface anesthesia, but that measured with von Frey filaments clearly increased after surface anesthesia, suggesting that the Randall-Selitto method measures the deep pressure withdrawal threshold

smaller probe (0.16 mm in diameter) gave a similar result. The pain threshold measured with probes having tip diameters of ≥1.6 mm did not change after surface anesthesia. Pressure transmission to deep tissues including muscle is expected to be blunted by thick subcutaneous fat, thus it is presumed that larger probes are needed to measure the muscle pain threshold with pressure. Because fat persons were not included in this experiment, it is probably safer to use tip diameters of >2 mm to measure the deep pain threshold in large populations including both fat and thin persons.

We also examined this issue in animals. We used a Randall-Selitto apparatus and the von Frey hair method. When the withdrawal threshold was measured by means of the Randall-Selitto apparatus with the commercially supplied probe, it was increased after surface anesthesia (EMLA cream; AstraZeneka, Grafenau, Switzerland), whereas if the tip diameter was increased to 2.3 mm the withdrawal threshold did not change even after surface anesthesia [64] (Fig. 2b, left panel). In contrast, the pain threshold measured with von Frey hairs with a tip diameter of 0.5 mm was clearly increased after surface anesthesia (Fig. 2b, right panel). Based on these results it is reasonable to conclude that the withdrawal threshold measured with a large probe reflects withdrawal threshold of the deep tissues.

To validate this result, we conducted a computer simulation of stress distribution in the tissue using a three-dimensional (3D) finite element analysis. We constructed a 3D finite element model of the forearm based on a magnetic resonance image of the human forearm [65]. The model consisted of 42 526 isotropic elastic solid elements. Two types of cylindrical pressure probe were created—one with a diameter of 0.5 mm and one of 10 mm. For each probe, pressure of 222.1 kPa was applied. The resultant stress in the muscle was one-tenth of that in the skin with the 0.5-mm probe and 0.4 with the 10-mm probe, which means that a larger probe can better transmit pressure to the deeper tissue. In conclusion, a larger probe can exert relatively greater stress to muscle with less stress to the skin.

Considering all these experimental and computer simulation data, we concluded that the withdrawal threshold of rats measured with a Randall-Selitto apparatus equipped with a large probe (2.3 mm diameter) represents the withdrawal threshold from deep tissues (probably the muscle). However, the Randall-Selitto threshold measured with a commercially supplied probe can also be used to measure the deep tissue threshold if the von Frey hair threshold is unchanged.

Method of Inducing DOMS

Downhill running has been used to induce DOMS in many studies; in our study, however, we electrically contracted the extensors—mainly the extensor digitorum longus muscle (EDL) of the lower hind leg—synchronously with stretching the muscle [49], so we could compare it with the unexercised side/parts and study the central processing of afferent information in the spinal cord. The gastrocnemius (GC) muscle can be also used for this purpose [48].

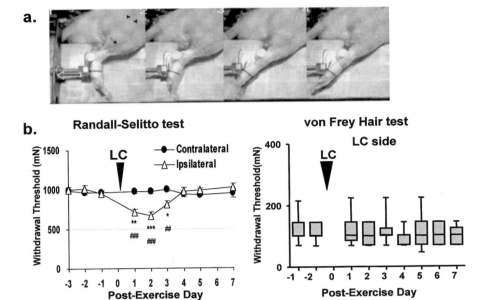

Fig. 2. Rat model of delayed-onset muscle soreness (DOMS). **a** Method applying lengthening contraction (*LC*) to an anesthetized rat. Photographs are arranged along the time course of one cycle of contraction from left to right. *Arrowheads* indicate a pair of needle electrodes for electrical stimulation of the common peroneal nerve. Contracted muscle can be recognized from the skin surface (*arrow*). While muscles are contracted, the hind paw was pulled by a motor to stretch the contracted muscles (large *arrow* in leftmost photograph). **b** Change in withdrawal thresholds measured with a Randall-Selitto apparatus (*left*) and von Frey hairs (*right*). Note that the Randall-Selitto threshold of the LC side decreased from 1 day to 3 days after exercise. Full recovery was seen 4 days after LC. Because there was no change in the von Frey hair threshold, it is concluded that the withdrawal threshold of deep tissues decreased after LC. (Modified from Taguchi et al. [61], with permission)

As shown in Fig. 2, a pair of needle electrodes, insulated except for the tips, were transcutaneously inserted near the common peroneal nerve, which innervates the EDL. Repetitive contraction of muscles (arrow in Fig. 2) including the EDL was induced by electrically stimulating the common peroneal nerve through these needle electrodes. The proper location of the electrodes was confirmed by the pattern of dorsiflexion of the ankle and extension of the middle three toes. The ankle joint was stretched (large arrow leftmost in Fig. 2a) synchronously with muscle contraction using a linearized servomotor. The contraction was repeated every 4 s for a total of 500 repetitions. Some rats (control group) received no current to their nerves, although the ankle joint was stretched as described above.

Mechanical Hyperalgesia After LC

Figure 2b shows the change in the mechanical withdrawal threshold measured with the Randall-Selitto apparatus (with a commercially supplied probe) after LC in young rats. Animals underwent LC on day 0. On day 1 the withdrawal threshold measured on the exercised extensors with the Randall-Selitto apparatus decreased slightly but significantly; and on day 2 it reached the bottom and remained decreased until day 3 after LC [49, 61]. On day 4 the threshold completely returned to the level prior to LC. The same result was obtained with a larger probe [49]. The withdrawal threshold of the contralateral side (no exercise) remained constant (unchanged) during the entire period of observation (Fig. 2b, left panel). There was no change in the mechanical withdrawal threshold of the skin over the exercised muscle measured with von Frey hairs (Fig. 2b, right). Based on these results we concluded that the deep tissues (probably the exercised muscle) of the exercised side were hyperalgesic to mechanical stimulation.

Itoh and Kawakita [48] used another method to detect mechanical hyperalgesia of the exercised muscle. They measured reflex EMG activity of the biceps femoris muscle in response to electrical stimulation to the GC muscle or to extension of the GC muscle. Under normal conditions, electrical stimulation or extension of the GC muscle induced little EMG activity in the biceps femoris muscle, whereas these manipulations induced a large amount of EMG activity 1 day after exercise, with the activity at its peak 2 days after exercise. These researchers also noticed that there was a sensitive point at the musculotendinous attachment area, which was located on the ropy band that was palpated with fingers through the skin as hardening of tissues (palpable taut band). They assumed this point to be an experimentally made trigger point. In our rat model, it was impossible to detect such a taut band because of the small size of the rat muscle.

c-Fos Study

To be sure that the observed decrease in the mechanical withdrawal threshold measured with the Randall-Selitto apparatus reflected mechanical hyperalgesia of exercised muscle, we examined c-Fos expression in the dorsal horn. C-Fos protein has been used as a neural marker of pain since Hunt et al. [66] reported that various kinds of noxious stimuli induced c-Fos protein in the superficial dorsal horn, which contains secondary neurons receiving nociceptive A-δ- and C-fiber inputs [67–70]. The characteristics of DOMS are hyperalgesia to compression- and movement-induced pain and the near absence of spontaneous pain. Therefore, in our c-Fos study the animals that underwent LC and the control group (which underwent stretching without contraction, Sham) groups were further divided into two subgroups with or without muscle compression. Compressive stimulation was applied to the exercised muscle through the skin on day 2 when the mechanical

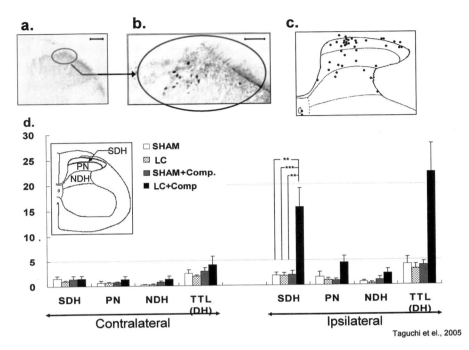

Fig. 3. Increased c-Fos expression after compression of the muscle 2 days after LC was observed in the superficial dorsal horn in the exercised side. **a, b** Sample photographs of the dorsal horn of a rat that received LC 2 days before and compression on the day of perfusion. **c** Camera lucida drawing of a representative section of rat that underwent LC and compression. **d** Number of c-Fos-immunoreactive neurons in various areas of the spinal dorsal horn at L4. *Comp.*, compression; *SDH*, superficial dorsal horn; *PN*, proprius nucleus; *NDH*, neck of dorsalhorn; *TTL (DH)*, total (dorsal horn). (Modified from Taguchi et al. [49], with permission)

hyperalgesia was at its peak (as shown in Fig. 2b, left). Compression with a force of 160 g was applied with a Randall-Selitto apparatus for 10 s, followed by a 20-s rest period. This was repeated for 30 min under anesthesia (pentobarbital 50 mg/kg i.p.). Two hours after the end of the muscle compression (or 2 days after the exercise session in groups without compression), the animals were deeply anesthetized and perfused with a fixative, and then L2–4 spinal coral was dissected and further processed for immunohistochemistry. Immunohistochemical staining of c-Fos was performed with a fixative according to the avidin-biotin peroxidase method of Hsu et al. [71].

C-Fos immunoreactivity was found in the nuclei of some neurons in the dorsal horn of the spinal cord (Fig. 3b). The number of c-Fos-immunoreactive (c-Fos-ir) neurons in the dorsal horn was small in the Sham, LC, and Sham+compression groups. Thus, compression used in this experiment did not activate the nociceptive pathway in anesthetized animals. In the LC+compression group, on the other hand, c-Fos immunoreactivity clearly increased, especially in the superficial dorsal horn corresponding to lamina I/II (Fig. 3b–d). Increased c-Fos expression was observed in the spinal cord at L2–4, but not at L5 or L6; the most prominent change was found in L4 [49]. This observation is in agreement with the previous report that most of the sensory neurons innervating the EDL muscle are located in the L4 dorsal root ganglion (DRG). The increased expression of c-Fos in the superficial dorsal horn in the LC+compression group was completely suppressed with morphine (10 mg/kg i.p.). These observations further support the supposition that the muscle is hyperalgesic 2 days after LC and suggest that this model can be useful for studying the neural mechanism of DOMS.

Change in DOMS in Aged Rats

The possibility that DOMS leads to more debilitating and chronic injury and that it results in chronic pain/hyperalgesia with plastic changes in the central nervous system, might be higher in elderly people and animals, as more severe muscle fiber damage [72] and slower recovery from damage have been reported in aged animals [73]. To determine whether aging produces any change in mechanical hyperalgesia after LC (DOMS), we applied LC to the lower hind leg extensors, including the EDL, in aged (81- and 130-week-old) rats as described above and examined the change in the mechanical withdrawal threshold of the exercised muscle with a Randall-Selitto apparatus and by c-Fos expression in the dorsal horn.

The baseline mechanical withdrawal threshold of these two groups did not differ from that of the young animals described in the previous section [74]. One day after LC the withdrawal threshold started to decrease in both age groups; however, the duration of the decreased withdrawal threshold was different: 81-week-old rats had their withdrawal threshold lowered for only 3 days after LC, similar to 7-week-old rats, whereas that of the 130-week-old rats remained lowered 2 days longer, showing delayed recovery in aged rats (Fig. 4a). Microinjuries in the muscle induced by LC are reported to induce inflammation, as introduced above; and recovery from inflammation in general is retarded in aged animals [75]. In addition, muscle in aged animals is reportedly more susceptible to injury during exercise [72, 76, 77], and the regeneration process of the muscle fibers is reportedly delayed in aged animals (see [78] for review). It is thus possible that this retarded recovery from inflammation and the retarded regeneration process are somehow related to the delayed recovery from mechanical hyperalgesia after exercise. If inflammation is not involved, as discussed in the previous section, delayed metabolism in aged animals may cause this delayed recovery. Alternatively, or additionally, increased central sensitization might be a cause of this increased duration of mechanical hyperalgesia, as suggested by the longer duration and larger area of punctate mechanical hyperalgesia after capsaicin treatment in elderly humans [79].

Induction of c-Fos expression in the spinal dorsal horn by compression of the muscle was examined in 130-week-old animals 3 days after LC (at the peak

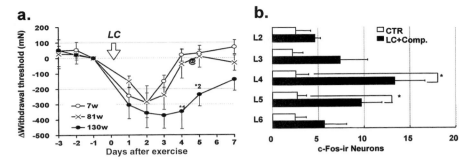

Fig. 4. Changes in mechanical hyperalgesia after LC and in c-Fos expression in aged rats. **a** Time course of mechanical hyperalgesia in three age groups (7, 81, and 130 weeks old) measured with the Randall-Selitto apparatus. Differences in withdrawal threshold from that on day −1 (1 day before LC loading) were plotted against days after exercise. **b** Segmental distribution of c-Fos expression in the superficial dorsal horn in 130-week-old rats. Note that increased expression was seen not only in L4 but also in L5, a wider distribution than in young rats (in L4 only). (From Taguchi et al. [61], with permission)

of mechanical hyperalgesia). Significantly larger numbers of c-Fos-ir-positive neurons were observed in the superficial dorsal horn of the animals that underwent compression of the muscle than in the control animals (no treatment). This increase was observed not only in L4 but also in L5 (Fig. 4b) [61]; this is a wider distribution than in young animals (in L4 only) described in the previous section and suggests a wider spread of mechanical hyperalgesia in aged animals.

Nociceptors in Muscle and Enhanced Mechanical Sensitivity of Thin-Fiber Receptors of Tender Muscle

It has been generally accepted that muscle pain is mediated by the activation of receptors with thin-fibers—groups III (A-δ) and IV (C)—which outnumber the thick myelinated fibers in muscular afferent nerves [80], although a possible contribution of large fiber afferents to soreness after LC has recently been suggested in human experiments [81, 82]. To determine the possible contribution of thin fibers to DOMS, we examined mechanical hyperalgesia after LC with the method reported by Taguchi et al. [49] in neonatally capsaicin (100mg/kg)-treated rats, in which most of the thin fibers were destroyed [83]. Vehicle-treated rats showed clear mechanical hyperalgesia after LC, similar to that reported previously [49]. In contrast, neonatally capsaicin-treated rats showed no mechanical hyperalgesia at all (unpublished observation), suggesting that thin fibers play a pivotal role in mechanical hyperalgesia in DOMS.

We then examined thin fiber activities in exercised animals. Muscle thin-fiber receptors are not only responsible for nociception, they are responsible for the early phase of exercise-induced hyperpnea, blood pressure changes, and tachycardia. It is not clear whether the fibers responsible for exercise-induced circulatory and respiratory changes are the same as nociceptors. Mechanical and chemical changes during exercise are thought to stimulate the receptors responsible for exercise-induced circulatory changes. Antagonists against substance P receptors [84] and ATP receptors [85, 86] suppress exercise-induced circulatory responses and afferent discharges, suggesting that these substances are involved in the exercise-induced pressor reflex. However, ATP can induce pain when infused into muscle [87].

The following observations suggest that the afferent system is different for muscle nociception and exercise-induced pressor reflex: Neonatal capsaicin treatment is known to destroy many small DRG neurons sensitive to capsaicin [83], which are thought to have thin fibers. Neonatally capsaicin-treated rats showed no mechanical hyperalgesia after LC (unpublished observation from our laboratory, described above), but they maintained a rather exaggerated exercise-induced pressor reflex [88]. However, it is reported that neonatally capsaicin-treated rats become exhausted faster during treadmill running than controls [89], so the possibility remains that this change in the muscle might be responsible for the absence of mechanical hyperalgesia after LC in neonatally capsaicin-treated rats. Further examination is needed.

Brief Review of Previous Works on Thin-Fiber Receptors in the Muscle

Muscle thin-fiber (A-δ and C) activities were first recorded in 1961 [90]. All of these receptors are classified as polymodal receptors in dogs; they respond to mechanical and thermal stimulation and to algesic substances such as bradykinin and hypertonic saline (Fig. 5) [18]. In contrast, thin-fiber afferents in cats were classified into four types: low-threshold mechanoreceptors, high-threshold mechanoreceptors, contraction-sensitive receptors, and thermosensitive C-fibers [91]. This classification was based on their sensitivity to various forms of mechanical stimulation (i.e., contraction, stretching, and pressure stimulation). Thermal stimulation in the nonnoxious range was done, but noxious thermal stimulation and algesic substances were not examined. Because later studies from the same group showed sensitivity to algesic chemicals of many mechanoreceptors (either high or low threshold with some preference in high-threshold ones), and 60% of thin-fiber receptors responded to capsaicin [92] (an agonist of heat-sensitive cation channels, TRPV1 [15]), many of these receptors seem to be a polymodal type. Evaluation of various types of stimulation is needed for classification.

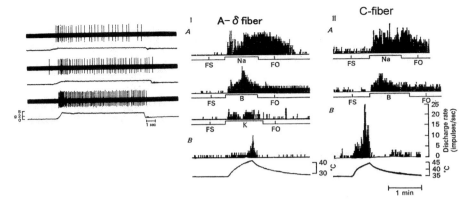

Fig. 5. Response pattern of canine A-δ- and C-fiber muscle polymodal receptors. **Left** Responses to graded mechanical stimulation of an A-δ fiber. **Middle** Responses to close-arterial injection of algesic substances and to heat of an A-δ fiber. **Right** Responses to algesic substances and to heat of a C-fiber. **Middle** and **right** panels include peristimulus histograms of discharges (1 bin is 1 s). *Na*, hypertonic saline (4.5%); *B*, bradykinin 1 μg/ml, *K*, high potassium solution (60 mM); *FS*, arresting blood flow; *FO*, reestablishing blood flow. (From Kumazawa and Mizumura [18], with permission)

Change in Inflammation

Few studies have been carried out on nociceptors in inflamed muscles. One study done on cats showed an increased percentage of group III (A-δ) fibers with spontaneous activity [29]. It also showed an increased percentage of group IV (C) fibers with thresholds in the moderate-pressure range. A similar observation was reported from the same group regarding rats [30].

More reports exist that show excitation and/or increased mechanical response by substances that may appear in inflamed muscles. Similar to cutaneous nociceptors, excitation was induced by bradykinin [18, 93], histamine [93], ATP [94, 95], serotonin [93], interleukin-6 (IL-6) [9], nerve growth factor (NGF) [9], and acid [92, 96], whereas sensitization to mechanical stimulation was induced by bradykinin [97] but not by NGF or IL-6 [9] by application for a short period.

Changes in DOMS

The existence of mechanical hyperalgesia in eccentrically contracted muscle was demonstrated by behavioral pain testing and c-Fos expression in the spinal cord (described earlier). To examine whether muscle thin-fiber activities, which transmit noxious information to the central nervous system (CNS), are changed with mechanical hyperalgesia after LC, we recorded single nerve fiber activity in vitro from EDL muscle–common peroneal nerve preparations and compared the mechanical sensitivities of thin fibers (conduction velocity <2.0m/s) between the control animals and those that underwent LC 2 days before [10].

Some thin fibers were spontaneously active, but there was no difference between the control and LC groups (0.13 ± 0.02 impulses (imp)/s in the controls and 0.09 ± 0.03 imp/s in the LC group). These values were not different from those reported previously from in vivo experiments [18, 98]; thus, the recorded receptors were considered to be in good condition. Receptive fields of the C-fiber sensory receptors were distributed all over the muscle, with a tendency to be concentrated near musculotendinous junctions (sample in Fig. 6a). The receptive field seemed to be more densely located near the musculotendinous junction in the LC preparations than in the controls. A ramp-shaped mechanical stimulation, linearly increasing from 0 to 198mN in 10s (shown in the inset of Fig. 6a) was applied by a servo-controlled mechanical stimulator with a flat circular tip (tip diameter about 1.6mm) to the point identified to be most sensitive.

Mechanical stimulation excited all of the fibers identified (33/33 fibers in the controls and 25/25 fibers in the LC group). A sample response of a fiber recorded in the control preparation is shown in Fig. 6a. The control preparation with no

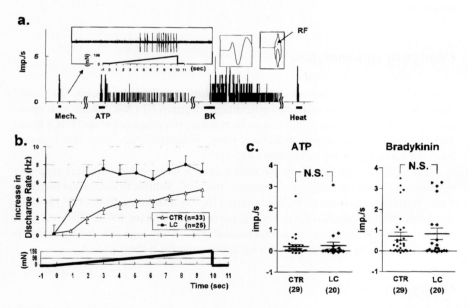

Fig. 6. Responses of rat muscle thin-fiber receptors to mechanical and algesic stimulation. **a** Sample recording from a C-fiber receptor in control rat EDL. *Mech.*, ramp mechanical stimulus (~196mN in 10s); *ATP*, 10mM adenosine 5'-triphosphate for 30s; *BK*, 10μM bradykinin for 60s. **b** Summary of the responses to ramp mechanical stimulation. *Triangles*, averaged response of receptors recorded from control preparations; *circles*, averaged response of receptors recorded from preparations 2 days after LC. **c** Response magnitudes to ATP (10mM) and bradykinin (10μM). Note that the mechanical response was clearly increased after exercise, but there was no difference in the responses to ATP and bradykinin

treatment showed, on average, an intensity-dependent increase of the discharge rate during a ramp-shaped mechanical stimulation (Fig. 6b, open triangles). On the other hand, the buildup of the response in the LC preparation was steeper than that in the control preparation (Fig. 6b, closed circles). Thus, the mechanical threshold in the LC preparations [median 38.2 mN, interquartile range (IQR); 26.8–55.8 mN, $n = 25$] was lower than that in the controls (median 65.4 mN, IQR; 46.6–122.0 mN, $n = 33$). In addition, the total number of evoked discharges during a ramp mechanical stimulation, as an index of the magnitude of the mechanical response, was significantly larger in the LC preparations than in the controls. This increased sensitivity of thin fibers is thought to be responsible for mechanical hyperalgesia after LC.

Other than increased sensitivity to mechanical stimulation, no difference was found in the responses to any other stimuli, algesic substances, acid, heat, or cold. For example, application of ATP activated 34.5% of fibers (10/29) in the controls and 50.0% (10/20) in the LC group (a sample response in Fig. 6a), which was not a statistically significant difference. The magnitude of the response to ATP was also not different between the controls and the LC group (Fig. 6c). Application of bradykinin induced activation of fibers in 65.5% of control fibers (19/29) and in 70.0% of the LC fibers (14/20), results that were not significantly different. The magnitude of the response to bradykinin was not different between the controls and the LC group (Fig. 6c). The fibers that were sensitive to ATP ($n = 10$ in the controls, $n = 10$ in the LC group) all responded to bradykinin.

Conclusions and Perspective

The DOMS model in rats was revealed to have purely mechanical hyperalgesia. Therefore, it is a good model for studying mechanical hyperalgesia in general, which is now one of the areas of most interest in nociception. In addition, pharmacological manipulation of this model can provide interesting information on the mechanism of developing DOMS. In future studies, muscle as well as muscular afferent neuron preparations, excised from DOMS rats and cultivated, can be used for these purposes. They provide a good opportunity to uncover the intracellular mechanisms of DOMS. Clinically more important is chronic muscle pain, and we are now preparing to develop a chronic muscle pain model based on DOMS with repeated cold stress application with the hope of finding a better way to treat chronic muscle pain.

References

1. Kellgren JH (1938) Observations on referred pain arising from muscle. Clin Sci 3:175–190.
2. Graven-Nielsen T, Arendt-Nielsen L, Svensson P, et al (1997) Stimulus-response functions in areas with experimentally induced referred muscle pain: a psychophysical study. Brain Res 744:121–128.

3. Graven-Nielsen T, Arendt-Nielsen L, Svensson M, et al (1997) Quantification of local and referred muscle pain in humans after sequential i.m. injections of hypertonic saline. Pain 69:111–117.
4. Graven-Nielsen T, Jansson Y, Segerdahl M, et al (2003) Experimental pain by ischaemic contractions compared with pain by intramuscular infusions of adenosine and hypertonic saline. Eur J Pain 7:93–102.
5. Babenko V, Graven-Nielsen T, Svensson P, et al (1999) Experimental human muscle pain induced by intramuscular injections of bradykinin, serotonin, and substance P. Eur J Pain 3:93–102.
6. O'Neill S, Manniche C, Graven-Nielsen T, et al (2007) Generalized deep-tissue hyperalgesia in patients with chronic low-back pain. Eur J Pain 11:415–420.
7. Bajaj P, Bajaj P, Madsen H, et al (2003) Endometriosis is associated with central sensitization: a psychophysical controlled study. J Pain 4:372–380.
8. Sorensen J, Graven-Nielsen T, Henriksson KG, et al (1998) Hyperexcitability in fibromyalgia. J Rheumatol 25:152–155.
9. Hoheisel U, Unger T, Mense S (2005) Excitatory and modulatory effects of inflammatory cytokines and neurotrophins on mechanosensitive group IV muscle afferents in the rat. Pain 114:168–176.
10. Taguchi T, Sato J, Mizumura K (2005) Augmented mechanical response of muscle thin-fiber sensory receptors recorded from rat muscle-nerve preparations in vitro after eccentric contraction. J Neurophysiol 94:2822–2831.
11. Mork H, Ashina M, Bendtsen L, et al (2003) Experimental muscle pain and tenderness following infusion of endogenous substances in humans. Eur J Pain 7:145–153.
12. Babenko V, Graven-Nielsen T, Svensson P, et al (1999) Experimental human muscle pain and muscular hyperalgesia induced by combinations of serotonin and bradykinin. Pain 82:1–8.
13. Cairns BE, Svensson P, Wang K, et al (2003) Activation of peripheral NMDA receptors contributes to human pain and rat afferent discharges evoked by injection of glutamate into the masseter muscle. J Neurophysiol 90:2098–2105.
14. Svensson P, Cairns BE, Wang K, et al (2003) Glutamate-evoked pain and mechanical allodynia in the human masseter muscle. Pain 101:221–227.
15. Caterina MJ, Schumacher MA, Tominaga M, et al (1997) The capsaicin receptor: a heat-activated ion channel in the pain pathway. Nature 389:816–824.
16. Witting N, Svensson P, Gottrup H, et al (2000) Intramuscular and intradermal injection of capsaicin: a comparison of local and referred pain. Pain 84:407–412.
17. Svensson P, Cairns BE, Wang K, et al (2003) Injection of nerve growth factor into human masseter muscle evokes long-lasting mechanical allodynia and hyperalgesia. Pain 104: 241–247.
18. Kumazawa T, Mizumura K (1977) Thin-fibre receptors responding to mechanical, chemical, and thermal stimulation in the skeletal muscle of the dog. J Physiol (Lond) 273:179–194.
19. Graven-Nielsen T, Arendt-Nielsen L, Mense S (2002) Thermosensitivity of muscle: high-intensity thermal stimulation of muscle tissue induces muscle pain in humans. J Physiol (Lond) 540:647–656.
20. Pan H-L, Longhurst JC, Eisenach JC, et al (1999) Role of protons in activation of cardiac sympathetic C-fibre afferents during ischaemia in cats. J Physiol (Lond) 518:857–866.
21. Meller ST, Gebhart GF (1992) A critical review of the afferent pathways and the potential chemical mediators involved in cardiac pain. Neuroscience 48:501–524.
22. Pan HL, Longhurst JC (1995) Lack of a role of adenosine in activation of ischemically sensitive cardiac sympathetic afferents. Am J Physiol 269:H106–H113.
23. Baker DG, Coleridge HM, Coleridge JC, et al (1980) Search for a cardiac nociceptor: stimulation by bradykinin of sympathetic afferent nerve endings in the heart of the cat. J Physiol (Lond) 306:519–536.
24. Vogt A, Vetterlein F, Ri HD, Schmidt G (1979) Excitation of afferent fibres in the cardiac sympathetic nerves induced by coronary occlusion and injection of bradykinin: the influence of acetylsalicylic acid and dipyron. Arch Int Pharmacodyn Ther 239:86–98.

25. Benson CJ, Eckert SP, Mccleskey EW (1999) Acid-evoked currents in cardiac sensory neurons: a possible mediator of myocardial ischemic sensation. Circ Res 84:921–928.
26. Molliver DC, Immke DC, Fierro L, et al (2005) ASIC3, an acid-sensing ion channel, is expressed in metaboreceptive sensory neurons. Mol Pain 1:35.
27. Naves LA, McCleskey EW (2005) An acid-sensing ion channel that detects ischemic pain. Braz J Med Biol Res 38:1561–1569.
28. Yagi J, Wenk HN, Naves LA, et al (2006) Sustained currents through ASIC3 ion channels at the modest pH changes that occur during myocardial ischemia. Circ Res 99:501–509.
29. Berberich P, Hoheisel U, Mense S (1988) Effects of a carrageenan-induced myositis on the discharge properties of group III and IV muscle receptors in the cat. J Neurophysiol 59: 1395–1409.
30. Diehl B, Hoheisel U, Mense S (1993) The influence of mechanical stimuli and of acetylsalicylic acid on the discharges of slowly conducting afferent units from normal and inflamed muscle in the rat. Exp Brain Res 92:431–440.
31. Ambalavanar R, Moritani M, Moutanni A, et al (2006) Deep tissue inflammation upregulates neuropeptides and evokes nociceptive behaviors which are modulated by a neuropeptide antagonist. Pain 120:53–68.
32. Asmussen E (1956) Observations on experimental muscular soreness. Acta Rheum Scand 2:109–116.
33. Armstrong RB (1984) Mechanisms of exercise-induced delayed onset muscular soreness: a brief review. Med Sci Sports Exerc 16:529–538.
34. Newham DJ (1988) The consequences of eccentric contractions and their relationship to delayed onset muscle pain. Eur J Appl Physiol 57:353–359.
35. Graven-Nielsen T, Arendt-Nielsen L (2003) Induction and assessment of muscle pain, referred pain, and muscular hyperalgesia. Curr Pain Headache Rep 7:443–451.
36. Armstrong RB, Oglive RW, Schwane JA (1983) Eccentric exercise-induced injury to rat skeletal muscle. J Appl Physiol 54:80–93.
37. McCully KK, Faulkner JA (1985) Injury to skeletal muscle fibers of mice following lengthening contractions. J Appl Physiol 59:119–126.
38. Friden J, Lieber RL (1998) Segmental muscle fiber lesions after repetitive eccentric contractions. Cell Tissue Res 293:165–171.
39. Newham DJ, McPhail G, Mills KR, et al (1983) Ultrastructural changes after concentric and eccentric contractions of human muscle. J Neurol Sci 61:109–122.
40. Ogilvie RW, Armstrong RB, Baird KE, et al (1988) Lesions in the rat soleus muscle following eccentrically biased exercise. Am J Anat 182:335–346.
41. Ostrowski K, Rohde T, Zacho M, et al (1998) Evidence that interleukin-6 is produced in human skeletal muscle during prolonged running. J Physiol (Lond) 508:949–953.
42. Blais C Jr, Adam A, Massicotte D, et al (1999) Increase in blood bradykinin concentration after eccentric weight-training exercise in men. J Appl Physiol 87:1197–1201.
43. Jones DA, Newham DJ, Round JM, et al (1986) Experimental human muscle damage: morphological changes in relation to other indices of damage. J Physiol (Lond) 375:435–448.
44. Schwane JA, Johnson SR, Vandenakker CB, et al (1983) Delayed-onset muscular soreness and plasma CPK and LDH activities after downhill running. Med Sci Sports Exerc 15: 51–56.
45. Malm C, Sjodin TL, Sjoberg B, et al (2004) Leukocytes, cytokines, growth factors and hormones in human skeletal muscle and blood after uphill or downhill running. J Physiol (Lond) 556:983–1000.
46. Cheung K, Hume P, Maxwell L (2003) Delayed onset muscle soreness: treatment strategies and performance factors. Sports Med 33:145–164.
47. Adams GR, Cheng DC, Haddad F, et al (2004) Skeletal muscle hypertrophy in response to isometric, lengthening, and shortening training bouts of equivalent duration. J Appl Physiol 96:1613–1618.
48. Itoh K, Kawakita K (2002) Effect of indomethacin on the development of eccentric exercise-induced localized sensitive region in the fascia of the rabbit. Jpn J Physiol 52:173–180.

49. Taguchi T, Matsuda T, Tamura R, et al (2005) Muscular mechanical hyperalgesia revealed by behavioural pain test and c-Fos expression in the spinal dorsal horn after eccentric contraction in rats. J Physiol (Lond) 564:259–268.
50. Schwane JA, Watrous BG, Johnson SR, et al (1983) Is lactic acid related to delayed-onset muscle soreness? Phys Sportsmed 11:124–131.
51. De Vries HA (1966) Quantitative electromyographic investigation of the spasm theory of muscle pain. Am J Phys Med 45:119–134.
52. Jones DA, Newham DJ, Clarkson PM (1987) Skeletal muscle stiffness and pain following eccentric exercise of the elbow flexors. Pain 30:233–242.
53. Friden J, Sjostrom M, Ekblom B (1983) Myofibrillar damage following intense eccentric exercise in man. Int J Sports Med 4:170–176.
54. Friden J, Lieber RL (1992) Structural and mechanical basis of exercise-induced muscle injury. Med Sci Sports Exerc 24:521–530.
55. Lapointe BM, Frenette J, Cote CH (2002) Lengthening contraction-induced inflammation is linked to secondary damage but devoid of neutrophil invasion. J Appl Physiol 92:1995–2004.
56. Matsuda T, Terazawa E, Mizumura K. (2007) Effect of nonsteroidal anti-inflammatory drugs (NSAIDS) and bradykinin (BK) receptor antagonists on mechanical hyperalgesia induced by exercise (DOMS). J Physiol Sci 57(suppl):S112 (abstract).
57. Vandenburgh HH, Shansky J, Solerssi R, et al (1995) Mechanical stimulation of skeletal muscle increases prostaglandin F_{2alpha} production, cyclooxygenase activity, and cell growth by a pertussis toxin sensitive mechanism. J Cell Physiol 163:285–294.
58. Wretman C, Lionikas A, Widegren U, et al (2001) Effects of concentric and eccentric contractions on phosphorylation of MAPK(erk1/2) and MAPK(p38) in isolated rat skeletal muscle. J Physiol (Lond) 535:155–164.
59. Haring HU, Tippmer S, Kellerer M, et al (1996) Modulation of insulin receptor signaling: potential mechanisms of a cross talk between bradykinin and the insulin receptor. Diabetes 45(suppl 1):S115–S119.
60. Inoue A, Iwasa M, Nishikura Y, et al (2006) The long-term exposure of rat cultured dorsal root ganglion cells to bradykinin induced the release of prostaglandin E_2 by the activation of cyclooxygenase-2. Neurosci Lett 401:242–247.
61. Taguchi T, Matsuda T, Mizumura K (2007) Change with age in muscular mechanical hyperalgesia after lengthening contraction in rats. Neurosci Res 57:331–338.
62. Fischer AA (1987) Pressure algometry over normal muscles: standard values, validity and reproducibility of pressure threshold. Pain 30:115–126.
63. Takahashi K, Taguchi T, Itoh K, et al (2005) Influence of surface anesthesia on the pressure pain threshold measured with different-sized probes. Somatosens Mot Res 22:299–305.
64. Nasu T, Terazawa E, Sato J, et al (2007) Randall-Selitto device measures the mechanical withdrawal thresholds of different tissues in rats depending on its probe diameter. J Physiol Sci 57(suppl):S112 (abstract).
65. Takahashi K, Taguchi T, Itoh K, et al (2004) Measurement of the muscle pain by a transcutaneous pressure: theoretical and experimental analyses. AbstractViewer/ Itinary Planner. Program No. 920.5 (abstract). Washington, DC, Society for Neuroscience.
66. Hunt SP, Pini A, Evan G (1987) Induction of c-fos-like protein in spinal cord neurons following sensory stimulation. Nature 328:632–634.
67. Cervero F, Connell LA (1984) Distribution of somatic and visceral primary afferent fibres within the thoracic spinal cord of the cat. J Comp Neurol 230:88–98.
68. Sugiura Y, Lee CL, Perl ER (1986) Central projections of identified, unmyelinated (C) afferent fibers innervating mammalian skin. Science 234:358–361.
69. Mizumura K, Sugiura Y, Kumazawa T (1993) Spinal termination patterns of canine identified A-δ and C spermatic polymodal receptors traced by intracellular labeling with phaseolus vulgaris leucoagglutinin. J Comp Neurol 335:460–468.
70. Ling LJ, Honda T, Shimada Y, et al (2003) Central projection of unmyelinated (C) primary afferent fibers from gastrocnemius muscle in the guinea pig. J Comp Neurol 461:140–150.

71. Hsu SM, Raine L, Fanger H (1981) Use of avidin-biotin-peroxidase complex (ABC) in immunoperoxidase techniques: a comparison between ABC and unlabeled antibody (PAP) procedures. J Histochem Cytochem 29:577–580.
72. Zerba E, Komorowski TE, Faulkner JA (1990) Free radical injury to skeletal muscles of young, adult, and old mice. Am J Physiol 258:C429–C435.
73. Brooks SV, Faulkner JA (1990) Contraction-induced injury: recovery of skeletal muscles in young and old mice. Am J Physiol 258:C436–C442.
74. Mizumura K, Taguchi T, Matsuda T, et al (2007) Change by aging in muscular mechanical hyperalgesia after lengthening contraction. Neurosci Res 55(suppl):S192 (abstract).
75. Kitagawa J, Kanda K, Sugiura M, et al (2005) Effect of chronic inflammation on dorsal horn nociceptive neurons in aged rats. J Neurophysiol 93:3594–3604.
76. McArdle A, Dillmann WH, Mestril R, et al (2004) Overexpression of HSP70 in mouse skeletal muscle protects against muscle damage and age-related muscle dysfunction. FASEB J 18:355–357.
77. McBride TA, Gorin FA, Carlsen RC (1995) Prolonged recovery and reduced adaptation in aged rat muscle following eccentric exercise. Mech Ageing Dev 83:185–200.
78. Close GL, Kayani A, Vasilaki A, et al (2005) Skeletal muscle damage with exercise and aging. Sports Med 35:413–427.
79. Zheng Z, Gibson SJ, Khalil Z, et al (2000) Age-related differences in the time course of capsaicin-induced hyperalgesia. Pain 85:51–58.
80. Stacey MJ (1969) Free nerve endings in skeletal muscle of the cat. J Anat 105:231–254.
81. Weerakkody NS, Whitehead NP, Canny BJ, et al (2001) Large-fiber mechanoreceptors contribute to muscle soreness after eccentric exercise. J Pain 2:209–219.
82. Weerakkody NS, Percival P, Hickey MW, et al (2003) Effects of local pressure and vibration on muscle pain from eccentric exercise and hypertonic saline. Pain 105:425–435.
83. Nagy JI, Iversen LL, Goedert M, et al (1983) Dose-dependent effects of capsaicin on primary sensory neurons in the neonatal rat. J Neurosci 3:399–406.
84. Kaufman MP, Kozlowski GP, Rybicki KJ (1985) Attenuation of the reflex pressor response to muscular contraction by a substance P antagonist. Brain Res 333:182–184.
85. Hanna RL, Kaufman MP (2003) Role played by purinergic receptors on muscle afferents in evoking the exercise pressor reflex. J Appl Physiol 94:1437–1445.
86. Kindig AE, Hayes SG, Kaufman MP (2007) Purinergic 2 receptor blockade prevents the responses of group IV afferents to post-contraction circulatory occlusion. J Physiol (Lond) 578:301–308
87. Mork H, Ashina M, Bendtsen L, et al (2003) Experimental muscle pain and tenderness following infusion of endogenous substances in humans. Eur J Pain 7:145–153.
88. Smith SA, Williams MA, Mitchell JH, et al (2005) The capsaicin-sensitive afferent neuron in skeletal muscle is abnormal in heart failure. Circulation 111:2056–2065.
89. Dousset E, Marqueste T, Decherchi P, et al (2004) Effects of neonatal capsaicin deafferentation on neuromuscular adjustments, performance, and afferent activities from adult tibialis anterior muscle during exercise. J Neurosci Res 76:734–741.
90. Iggo A (1961) Non-myelinated afferent fibers from mammalian skeletal muscle. J Physiol (Lond) 155:52–53.
91. Mense S, Meyer H (1985) Different types of slowly conducting afferent units in cat skeletal muscle and tendon. J Physiol (Lond) 363:403–417.
92. Hoheisel U, Reinohl J, Unger T, et al (2004) Acidic pH and capsaicin activate mechanosensitive group IV muscle receptors in the rat. Pain 110:149–157.
93. Fock S, Mense S (1976) Excitatory effects of 5-hydroxytryptamine, histamine and potassium ions on muscular group IV afferent units: a comparison with bradykinin. Brain Res 105:459–469.
94. Reinohl J, Hoheisel U, Unger T, et al (2003) Adenosine triphosphate as a stimulant for nociceptive and non-nociceptive muscle group IV receptors in the rat. Neurosci Lett 338:25–28.

95. Hanna RL, Kaufman MP (2004) Activation of thin-fiber muscle afferents by a P2X agonist in cats. J Appl Physiol 96:1166–1169.
96. Sinoway LI, Hill JM, Pickar JG, et al (1993) Effects of contraction and lactic acid on the discharge of group III muscle afferents in cats. J Neurophysiol 69:1053–1059.
97. Mense S, Meyer H (1988) Bradykinin-induced modulation of the response behaviour of different types of feline group III and IV muscle receptors. J Physiol (Lond) 398:49–63.
98. Franz M, Mense S (1975) Muscle receptors with group IV afferent fibres responding to application of bradykinin. Brain Res 92:369–383.

ASIC3 and Muscle Pain

Chih-Cheng Chen

Summary

Highly sensitive acid-sensing ion channel 3 (ASIC3) is predominantly distributed in sensory neurons. Moderate acidification (pH 6.7–7.3) can open ASIC3 with sustained inward current, which is suitable for persistent pain sensation evoked by acidosis in arthritis, muscular ischemia, and cancer, for example. Mice with null mutation of ASIC3 show enhanced responses to cutaneous pain stimuli, including heat, pressure, and acetic acid. Also, the hyperalgesia induced by intradermal inflammation is not changed or augmented. This evidence argues against ASIC3 triggering cutaneous pain. In contrast, muscle pain associated with acidosis requires ASIC3 for the development of chronic mechanical hyperalgesia. Therefore, ASIC3 in muscle afferents seems to act as a peripheral initiator of muscle pain associated with acidosis.

Key words ASIC3, Acid-evoked pain, Hyperalgesia, DRG, Muscle afferent

Introduction

Pain with musculoskeletal origin is common in humans. This pain is often felt as "acidic pain," as described by most people. The acidic pain sensation may be initiated by the decreased pH following inflammation, ischemia, hematoma, and isometric exercise. Based on electrophysiological studies, low pH activates small-diameter muscle afferent neurons in vitro and unmyelinated (group IV) muscle afferents in vivo in rodent models [1, 2]. In humans, both constant infusion of pH 5.2 phosphate buffer into muscle and ischemic contractions produce pain [3]. However, research into the cellular and molecular mechanisms of acidic muscle pain is rare. Recently, a series of studies demonstrated that acid-sensing ion channel 3 (ASIC3) in muscle afferents is the key acid sensor initiating long-lasting

---

Institute of Biomedical Sciences, Academia Sinica, 128 Yen-Chiu-Yuan Road, Section 2, Taipei 115, Taiwan

hyperalgesia in muscle [4–6]. The ASIC3 signaling pathway seems to shed light on clinical treatment for chronic muscle-induced hyperalgesia.

Acid and Pain

Sensing tissue acidosis is an important function for primary sensory neurons in detecting inflammatory, ischemic, and metabolic states. In response to extracellular acidification, sensory neurons exhibit a variety of depolarizing currents [7, 8]. The ion channels responsible for these currents include capsaicin receptor TRPV1, acid-sensing ion channels (ASICs), and the TWIK-related acid-sensing K^+ channel (TASK) [9]. Although nociceptive neurons express all these proton-gated ion channels, human study results strongly suggest that activation of ASICs likely plays a role in acid-evoked pain [10, 11]. Only in highly acidic conditions is TRPV1 involved in pain perception [11].

Acid-Sensing Ion Channels

Acid-sensing ion channels are members of the epithelium sodium channel (ENaC)/degenerin family, which have two membrane-spanning regions with intracellular N and C termini and a large extracellular loop [12]. The channels were first cloned from brain tissues by use of an expressed sequence tag (GenBank accession no. Z45660), with homology to ENaC and degenerin, so they were named brain Na^+ channels (e.g., BNC1) or mammalian degenerin (MDEG) [13, 14]. No function was reported for BNC1 and its homologue BNC2 until Lazdunski's group identified the proton ligand for BNaC2 and renamed the protein acid-sensing ion channel 1 (ASIC1) [15, 16]. The heterologously expressed ASIC1 channel is permeable to Na^+, Ca^{2+}, and K^+ ($Na^+ > Ca^{2+} > K^+$) and mediates an inward current when external pH is decreased rapidly from pH 7.4 to less than pH 6.9. The pioneer work encouraged further cloning of its splice variant ASIC1b [17] and other homologues: ASIC2a (origin as BNaC1), ASIC2b (splice variant to ASIC2a) [18], ASIC3 (DRASIC) [19], ASIC4 (SPASIC) [20], and ASIC5 (BLINaC) [21]. Like ASIC1 (now renamed ASIC1a), ASIC1b, ASIC2a, and ASIC3 behave as proton-gated ion channels, but each has distinct pH sensitivity and channel kinetics. Heterologous expression of individual ASIC1a, AIC1b, ASIC2a, and ASIC3 generates a transient proton-gated sodium current that is inhibited by amiloride. ASIC2b is known as a modulatory subunit to modify the proton-gated property of ASIC channels. ASIC4 and ASIC5 have no known function.

All these ASIC subunits are found in sensory neurons, although ASIC2a, ASIC4, and ASIC5 are detectable only by the polymerase chain reaction or in situ hybridization. The distribution of ASIC1a is restricted to peptidergic small-diameter dorsal root ganglion (DRG) neurons, whereas ASIC1b, ASIC2b, and

ASIC3 are distributed in a whole range of cell sizes of DRG neurons [1, 16, 21]. In DRG neurons, multiple ASIC subunits assemble to form a functional channel, which is either homomeric or heteromeric [22–24].

Electrophysiological Properties of ASIC3

ASIC3 is the most sensitive acid-sensing ion channel, with the pH sensitivity near the physiological condition (pH 6.7–7.3) [23, 25]. In response to acid stimulation, heterologously expressed ASIC3 mediates a transient inward current and a sustained inward current. In contrast, other ASIC subunits generate only a proton-evoked transient inward current when expressed in cell lines or oocytes [12]. Because the fast inactivation of an ASIC-mediated current cannot account for the prolonged sensation of pain associated with tissue acidosis, the ASIC3-mediated sustained current is much scrutinized. ASIC3 is the only proton-gated ion channel that generates a sustained current in response to moderate acidification. The ASIC3-mediated sustained current has been demonstrated as being essential for firing action potential in native sensory neurons [26]. Therefore, the electrophysiological properties have raised the likelihood that ASIC3 mediates pain associated with cardiac ischemia, muscle ischemia, or inflammation. Moreover, the ASIC inhibitor amiloride has been demonstrated to attenuate acid-induced pain in human skin [9, 10]. Because amiloride not only blocks ASIC3 but also inhibits other ASIC channels and ENaC, as well as many other channels, the roles for ASIC3 in pain sensation have yet to be verified by use of more specific blockers, such as the sea anemone peptide APETx2 [27].

ASIC3 Knockout Mice

To understand the roles of ASIC3 in pain and physiological functions, several groups have generated mice with a null mutation of ASIC3 [28–30]. However, the results of pain features in ASIC3 knockout mice are controversial. In a skin nerve preparation, the deletion of ASIC3 in mice resulted in a blunted response to acid in a subpopulation of C-fibers but enhanced the activity of rapidly adapting mechanoreceptors [28]. However, nociceptive behaviors after acid injection in the paw are not altered. In contrast, ASIC3 knockout mice showed enhanced mechanical hyperalgesia following carrageenan-induced paw inflammation. Moreover, ASIC3 was responsible for hyperalgesic effects with high-intensity pain stimuli in tests involving a hot plate (heat hyperalgesia), tail pressure (mechanical hyperalgesia), and intraperitoneal acetic acid (chemical hyperalgesia) treatment [29]. Similarly, dysfunction of ASIC3 in a dominant-negative ASIC3 transgenic mouse model led to hyperalgesic effects in many nociceptive assays [31]. Taken together, these studies seem to support the role of ASIC3 as being not specific to acid-evoked pain but, rather, having a more diversified sensory function. Although ASIC3 knockout and dominant-negative models all show unexpected increased sensitivity to various pain stimuli, ASIC3 is required for the development of secondary mechanical hyperalgesia induced by acid injected into skeletal muscle [5, 28, 31]. This discrepancy of ASIC3 function in nociception may be due to predominant innervation of ASIC3-positive C-fibers into muscle tissue but not cutaneous tissue [1].

Role of ASIC3 in Intramuscular Acid-Induced Chronic Hyperalgesia

The best study to demonstrate the role of ASIC3 in acid-evoked pain was in a muscle-originated pain model. Sluka's group developed intramuscular acid-induced chronic hyperalgesia in rodents [4]. In this model, repeated injections of acidic saline into one gastrocnemius muscle produced bilateral, long-lasting mechanical hypersensitivity of the paw without associated tissue damage. Following the first intramuscular acid injection, the withdrawal threshold to von Frey filaments (mechanical stimulation) was decreased at 4 h and returned to baseline by 24 h. This mechanical hyperalgesia was observed only in the ipsilateral hindpaw, not the contralateral paw. Following the second acid injection (2 or 5 days after the first injection), both ipsilateral and contralateral hindpaws developed mechanical hyperalgesia, which persisted for 4 weeks. Interestingly, this model did not develop thermal hyperalgesia. The intramuscular acid-induced hyperalgesia is maintained by changes in the central nervous system because it is reversed by intraspinal administration of μ- or δ-opioid receptor agonists [32] or *N*-methyl-d-aspartate (NMDA) or non-NMDA ionotropic glutamate receptor antagonists [33]. Parallel to this, spinal-wide dynamic neurons are sensitized after the second acid injection [5]. This acid-induced chronic mechanical hyperalgesia is initiated by activation of ASIC3 in muscle afferents because the hyperalgesia is normally developed in ASIC1 knockout mice but abolished in ASIC3 knockout mice [5]. Intramuscular injection of neurotrophin-3 can reverse the acid-induced chronic mechanical hyperalgesia [34]. Interestingly, ASIC3-positive muscle afferent neurons co-express with TrkC, tyrosine kinase receptors with high affinity for neutrophin-3 [1].

ASIC3 in Hyperalgesia Associated with Muscle Inflammation

Chronic musculoskeletal pain often results from repetitive strain injury and tendonitis, which are associated with muscle inflammation. Although inflammation causes decreased pH, the intramuscular acid-induced hyperalgesia is a noninflammatory muscle pain model. In contrast to acid-induced hyperalgesia, a single intramuscular injection of carrageenan causes muscle inflammation, and both heat and mechanical hyperalgesia develop [35]. This muscle inflammation-induced hyperalgesia is initiated on the ipsilateral side and spreads to the contralateral side 1–2 weeks after injection. Null mutation of ASIC3 attenuates the inflammation-induced mechanical but not heat hyperalgesia in mice [6]. Interestingly, injection of an ASIC3-encoding virus into muscle but not skin can restore the muscle inflammation-induced mechanical hyperalgesia in ASIC3 knockout mice. However, cutaneous inflammation by carrageenan induces or even enhances hyperalgesia in ASIC3 knockout mice or dominant-negative mutants [28, 29, 31], although cutaneous inflammation increases the transcription levels of ASIC3 in the DRG [36]. Therefore, ASIC3 in muscle afferents is related to nociception and is essential for the development of mechanical hyperalgesia associated with muscle inflammation.

ASIC3 in Ischemic Muscle Pain

Aside from the intramuscular acid injection and muscle inflammation, exercising ischemic muscle causes pain [3]. Such muscle pain is usually felt as acidic pain and can be experienced by running or swimming at a sprint speed for less than 1 min. The ischemic pain does not occur in the skin when blood flow is constrained for thermoregulation or for flight response. The channel property and distribution of ASIC3 suggests a role in ischemic muscle pain. First, the ASIC3 channel can be opened at within 0.5 unit of pH changes, the range at which pH decreases during muscle ischemia [3, 26]. Second, lactate (in millimoles) enhances ASIC3 sensitivity to protons [25, 37]. Therefore, ASIC3 is a suitable sensor for lactic acidosis when muscle is using anaerobic metabolism. Third, ASIC3-positive sensory neurons are enriched in small muscle afferents but not in small skin afferents [1]. Despite these special features of ASIC3, the real role of ASIC3 in ischemic pain has not been verified in an in vivo model. With the advantage of ASIC3 knockout mice, ischemic muscle pain models in mice can be useful for determining the role of ASIC3.

ASIC3-Expressing Muscle Afferent Neurons

ASIC3 is unique in its predominant expression in sensory neurons [17, 19]. Both immunostaining and in situ hybridization have revealed ASIC3 in small-diameter and medium- to large-diameter DRG neurons [1, 38]. We as well as others have used electrophysiological measurements to demonstrate ASIC3 preferentially expressed in medium- to large-diameter neurons and only in a small portion of small-diameter neurons [22]. Nevertheless, the distribution of ASIC3 in various-size neurons implies that the channels may be involved in diverse sensory modes. Small-diameter sensory neurons refer to the unmyelinated C-fiber neurons, most of which are nociceptive neurons [39]. Medium- to large-diameter sensory neurons refer to myelinated $A\delta$ or $A\alpha/\beta$ fiber neurons, which are nociceptive or nonnociceptive (mechanical or proprioceptive) neurons, respectively [40]. With retrograde labeling of rat skeletal muscle, more than half of the small muscle afferent neurons (18/32) expressed an ASIC3-like current [5]. In contrast, only 11% (4/37) of small skin afferent neurons expressed the ASIC3-like current. In the mouse, one-half (10/20) of muscle afferent neurons responded to pH 5.0 with an inward current, but only 15% (3/20) expressed an ASIC3-like biphasic inward current [5]. ASIC3-positive muscle afferent neurons were of medium size and appeared to be Aδ fiber neurons [1]. Most of the ASIC3-positive Aδ muscle afferent neurons co-expressed with TRPV1 and CGRP and innervated the outer layer of small arteries in muscle. Such anatomical organization supports ASIC3-positive muscle afferents detecting lactic acidosis and ischemic muscle pain. Upon activation, nociceptive neurons release CGRP to increase blood flood through vasodilation [41]. With co-expression of CGRP and ASIC3, these muscle arterial afferents can function together as sensor and effector to respond to and compensate for muscle ischemia.

Future Perspectives

Chronic musculoskeletal pain, which is common in our society, is difficult to treat; and the underlying mechanism is still largely unknown. Recent studies demonstrate that the ASIC3 of muscle afferents may be the critical peripheral initiator of mechanical hyperalgesia resulting from muscle insult. Chronic mechanical hyperalgesia induced by repeated intramuscular acid injection or muscle inflammation requires ASIC3. These two muscle pain models clearly link ASIC3 to acidic pain. However, whether ASIC3 plays a role in isometric exercise-induced and/or ischemic muscle pain has yet to be determined. In addition, a muscle soreness model with eccentric contraction would be worthy for testing in ASIC3 knockout mice [42]. Because of the importance of ASIC3 in acid-evoked pain originating from muscle, knowing the underlying sensory pathways to the central nervous system is important. Tools for new brain imaging and systemic electrophysiology in mice are expected to disclose the ASIC3-mediated pain signal pathways and the relevance of the channel to clinical chronic musculoskeletal pain such as fibromyalgia, myofacial pain syndrome, and complex regional pain syndromes.

References

1. Molliver DC, Immke DC, Fierro L, et al (2005) ASIC3, an acid-sensing ion channel, is expressed in metaboreceptive sensory neurons. Mol Pain 1:35.
2. Hoheisel U, Reinohl J, Unger T, et al (2004) Acidic pH and capsaicin activate mechanosensitive group IV muscle receptors in rat. Pain 110:149–157.
3. Issberner U, Reeh PW, Steen KH (1996) Pain due to tissue acidosis: a mechanism for inflammatory and ischemic myalgia? Neurosci Lett 208:191–194.

4. Sluka KA, Kalra A, Moore SA (2001) Unilateral intramuscular injections of acidic saline produce a bilateral, long-lasting hyperalgesia. Muscle Nerve 24:37–46.
5. Sluka KA, Price MP, Breese NM, et al (2003) Chronic hyperalgesia induced by repeated acid injections in muscle is abolished by the loss of ASIC3, but not ASIC1. Pain 106:229–239.
6. Sluka KA, Radhakrishnan R, Benson CJ, et al (2006) ASIC3 in muscle mediates mechanical, but not heat, hyperalgesia associated with muscle inflammation. Pain 129:102–112.
7. Krishtal OA, Pidoplichko VI (1981) A receptor for protons in the membrane of sensory neurons may participate in nocicpetion. Neuroscience 6:2599–2601.
8. Bevan S, Yeat J (1991) Protons activate a cation conductance in a subpopulation of rat dorsal root ganglion neurons. J Physiol 433:145–161.
9. Reeh PW, Kress M (2001) Molecular physiology of proton transduction in nociceptors. Curr Opin Pharmacol 1:45–51.
10. Ugawa S, Ueda T, Ishida Y, et al (2002) Amiloride-blockable acid-sensing ion channels are leading acid sensors expressed in human nociceptors. J Clin Invest 110:1185–1190.
11. Jones NG, Slater R, Cadiou H, et al (2004) Acid-induced pain and its modulation in humans. J Neurosci 24:10974–10979.
12. Waldmann R, Lazdunski M (1998) H^+-gated cation channels: neuronal acid sensors in the NaC/DEG family of ion channels. Curr Opin Neurobiol 8:418–424.
13. Prices MP, Synder PM, Welsh MJ (1996) Cloning and expression of a novel human brain Na^+ channel. J Biol Chem 271:7879–7882.
14. Waldmann R, Champigny G, Voilley N, et al (1996) The mammalian degenerin MDEG, an amiloride-sensitive cation channel activated by mutations causing neurodegeneration in *Caenorhabditis elegans*. J Biol Chem 271:10433–10436.
15. Garcia-Anoveros J, Drfler B, Neville-Golden J, et al (1997) BNaC1 and BNaC2 constitute a new family of human neuronal sodium channels related to degenerins and epithelium sodium channels. Proc Natl Acad Sci USA 94:1459–1464.
16. Waldmann R, Champigny G, Bassilana F, et al (1997) A proton-gated cation channel involved in acid-sensing. Nature 386:173–177.
17. Chen CC, England S, Akopian AN, et al (1998) A sensory neuron-specific, proton-gated ion channel. Proc Natl Acad Sci USA 95:10240–10245.
18. Lingueglia E, de Weille JR, Bassilana F, et al (1997) A modulatory subunit of acid sensing ion channels in brain and dorsal root ganglion cells. J Biol Chem 272:29778–29783.
19. Waldmann R, Bassilana F, de Weille J, et al (1997) Molecular cloning of a non-inactivating proton-gated Na+ channel specific for sensory neurons. J Biol Chem 272:20975–20978.
20. Akopian AN, Chen CC, Ding YN, et al (2000) A new member of the acid-sensing ion channel family. Neuroreport 11:2217–2222.
21. Sakai H, Lingueglia E, Champingny G, et al (1999) Cloning and functional expression of a novel degenerin-like Na^+ channel gene in mammals. J Physiol 519:323–333.
22. Alvarez de la Rosa D, Zhang P, Shao D, et al (2002) Functional implications of the localization and activity of acid-sensitive channels in rat peripheral nervous system. Proc Natl Acad Sci USA 99:2326–2331.
23. Sutherland SP, Benson CJ, Adelman JP, et al (2001) Acid-sensing ion channel 3 matches the acid-gated current in cardiac ischemia-sensing neurons. Proc Natl Acad Sci USA 98: 711–716.
24. Benson CJ, Xie J, Wemmie JA, et al (2002) Heteromultimers of DEG/ENaC subunits for, H^+-gated channels in mouse sensory neurons. Proc Natl Acad Sci USA 99:2338–2343.
25. Immke DC, McCleskey EW (2001) Lactate enhances the acid-sensing Na^+ channel on ischemia-sensing neurons. Nat Neurosci 4:869–870.
26. Yagi J, Wenk HN, Naves LA, et al (2006) Sustained currents through ASIC3 ion channels at the modest pH changes that occur during myocardial ischemia. Circ Res 99:501–509.
27. Diochot S, Baron A, Rash LD, et al (2004) A new sea anemone peptide, APETx2, inhibits ASIC3, a major acid-sensitive channel in sensory neurons. EMBO J 23:1516–1525.
28. Price MP, McIlwrath SL, Xie J, et al (2001) The DRASIC cation channel contributes to the detection of cutaneous touch and acid stimuli in mice. Neuron 32:1071–1083.

29. Chen CC, Zimmer A, Sun WH, et al (2002) A role for ASIC3 in the modulation of high-intensity pain stimuli. Proc Natl Acad Sci USA 99:8992–8997.
30. Drew LJ, Rohrer DK, Price MP, et al (2004) Acid-sensing ion channels ASIC2 and ASIC3 do not contribute to mechanically activated currents in mammalian sensory neurons. J Physiol 556:691–710.
31. Mogil JS, Breese NM, Witty MF, et al (2005) Transgenic expression of a dominant-negative ASIC3 subunit leads to increase sensitivity to mechanical and inflammatory stimuli. J Neurosci 25:9893–9901.
32. Sluka KA, Rohlwiing JJ, Bussey RA, et al (2002) Chronic muscle pain induced by repeated acid injection is reversed by spinally administered µ-, and δ-, but not κ-, opioid receptor agonists. J Pharmacol Exp Ther 302:1146–1150.
33. Skyba DA, Kiing EW, Sluka KA, et al (2002) Effects of NMDA and non-NMDA ionotropic glutamate receptor antagonists on the development and maintenance of hyperalgesia induced by repeated intramuscular injection of acid saline. Pain 98:69–78.
34. Gandhi R, Ryals JM, Wright DE (2004) Neurotrophin-3 reverses chronic mechanical hyperalgesia induced by intramuscular acid injection. J Neurosci 24:9405–9413.
35. Radhakrishnan R, Moore SA, Sluka KA (2003) Unilateral carrageenan injection into muscle or joint induces chronic bilateral hyperalgesia in rats. Pain 104:567–577.
36. Voilley N, de Weille J, Mamet J, et al (2001) Nonsteroid anti-inflammatory drugs inhibit both the activity and the inflammation-induced expression of acid-sensing ion channels in nociceptors. J Neurosci 21:8026–8033.
37. Immke DC, McCleskey EW (2003) Protons open acid-sensing ion channels by catalyzing relief of Ca^{2+} blockade. Neuron 37:75–84.
38. Ugawa S, Ueda T, Yamamura H, et al (2005) In situ hybridization evidence for the coexistence of ASIC and TRPV1 within rat single sensory neurons. Mol Brain Res 136:125–133.
39. Harper AA, Lawson SN (1985) Conduction velocity is related to morphological cell type in rat dorsal root ganglion neurons. J Physiol 359:31–46.
40. Djouhri L, Laeson SN (2004) Abeta-fiber nociceptive primary afferent neurons: a review of incidence and properties in relation to other afferent A-fiber neurons in mammals. Brain Res Brain Res Rev 46:131–145.
41. Willis WD Jr (1999) Dorsal root potentials and dorsal root reflexes: a double-edged sword. Exp Brain Res 124:395–421.
42. Taguchi T, Matsuda T, Tamura R, et al (2005) Muscular mechanical hyperalgesia revealed by behavioural pain test and c-Fos expression in the spinal dorsal horn after eccentric contraction in rats. J Physiol 564:259–268.

Tail Region of the Primary Somatosensory Cortex and Its Relation to Pain Function

Chen-Tung Yen[1,2] and Ren-Shiang Chen[3]

Summary

In the present study, electrophysiological mapping methods were used to estimate the size of the tail representation area of the primary somatosensory cortex (SI) of the rat. Using a half-maximal evoked potential method and multiunit recording method, we estimated that the SI tail area was 0.51 and 0.78 mm^2, respectively. A dissector method was used to estimate the neuronal densities. There was, on average, 84 829 neurons/mm^3 and 117 750 neurons under 1 mm^2 of cortical area in the tail area of the SI. Therefore, there are about 94 000 neurons in the estimated 0.8 mm^2 of the SI that are involved in processing sensory signals from the tail. Anteroposteriorly oriented, evenly spaced 16-channel microwires were chronically implanted in the frontoparietooccipital cortex that was centered on the SI. Thereafter, evoked field potentials were used to estimate the change in the size of the tail area with two modalities—pain and touch—under two states: anesthetized and conscious. No significant difference was found between the size of the tail area when tactile and noxious stimulations were used. However, the number of tail responsive channels showed a significant increase when the rat was awake and behaving.

Key words Dissector method, Neuronal density, Pain, Primary somatosensory cortex, Tail

Introduction

The sensorimotor system of the rat tail has been a useful model system in pain research, such as with the tail flick test [1]. This system has many advantages to recommend as a model somatosensory system. The tail of the rat is long and

---

[1]Institute of Zoology, National Taiwan University, 1 Roosevelt Road, Section 4, Taipei 106, Taiwan

[2]Research Center of Brain and Oral Science, Kanagawa Dental College, Kanagawa, Japan

[3]Department of Zoology, National Taiwan University, Taipei, Taiwan

geometrically simple, and it is thus an easy target for applying specific stimuli. Second, a complete set of receptors of various somatosensory modalities, such as touch, temperature, and pain, are present in the tail. Third, because spinal roots of the tail neurons are long and thinly sheathed [2], selective activation and recording of A- or C-fibers is easier [3]. Finally, the central representation areas of the tail are relatively small [4–6], so the detailed neural circuitry of this system is easier to reach.

The somatosensory system of mammals is hierarchically organized with components in the spinal cord, medulla, thalamus, and cortex. To model any part of this system realistically, it is necessary to know each of the components involved and the fine circuitry of the intra- and interconnections. The primary somatosensory cortex (SI) is an important component in the middle of this multilayered organization. Therefore, the purpose of the present study was to estimate the number of neurons in the SI of the rat that might be involved in processing tactile and nociceptive inputs from the tail.

To obtain an estimation of total neuron number in a brain area for a given function, three sets of data are needed: the location and size of the brain area devoted to the given function and the neuronal density in this location. The tail region of the SI of the rat has been mapped many times using electrophysiological techniques [4, 6, 7]. A detailed ratunculus is shown in each article. Because these maps are composites of a great many rats, they are of little use in quantitative size estimation. The first part of the present study used unit recording and evoked field potential mapping techniques to obtain an estimate of the size of the tail area in the SI of individual rats.

In our experience, cortices that have been mapped with microelectrode penetration are swollen and distorted to varying degrees. They are usually unsuitable for morphometric measurements. The second purpose of the present study, therefore, was to find a corresponding morphological landmark for identifying the tail area of the SI in serial histological sections. This was accomplished by looking for a correspondence between the medialmost tip of layer IV (the granular layer) with the tail tip region mapped using electrophysiological recordings.

To estimate the neuronal density in the medial-tip region of the SI and adjacent areas, the dissector method [8] was used. The tail area of the SI has an estimated 117 000 neurons/mm^2, which when multiplied by the estimated size of 0.8 mm^2 gives approximately 94 000 neurons in the SI that might be involved in processing inputs from the tail.

To probe the relevance of the above data to pain processing, we used radiant laser heat pulses as a specific noxious stimulus on the tail and made recordings from the tail area of the SI in chronically instrumented rats. Our results showed that the tail area of the SI (determined by tactile stimulation while a rat was anesthetized) closely corresponded to the largest nociceptive responsive channel. The size of the nociceptive responsive area, however, greatly expanded when the rat was conscious and behaving. How this dynamically changing cortical representation pattern relates to pain function is discussed.

Original Experiments

Adult rats weighing 270–450 g were used. A preliminary survey of the innervation of the tail by gross dissection and electrophysiological mapping of the dermatome of the segmental dorsal root was undertaken. The tail of the rat is innervated by sacrococcygeal segments of the spinal cord (Fig. 1). The four sacral segmental nerves and the two coccygeal segmental nerves form a complicated caudal plexus to supply the two pairs of caudal nerves going to the tail (Fig. 2).

Three experiments were performed: (1) mapping experiments to determine the size of the tail area of the SI; (2) stereological experiments to determine the neuronal densities in the tail area of the SI; and (3) a comparison of the size of the tactile and noxious responsive areas in the tail area of the SI in anesthetized rats and in conscious, behaving rats.

Electrophysiological Mapping Experiment

Rats were anesthetized with an intraperitoneal injection of sodium pentobarbital (50 mg/kg), and the trachea, right femoral artery, and vein were cannulated. The trachea cannula was to ensure adequate ventilation. The animals breathed spontaneously during the entire recording period. The arterial blood pressure was measured through a pressure sensor connected to the arterial cannula. The venous cannula

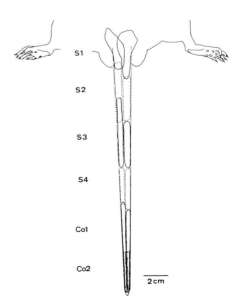

Fig. 1. Dermatome of sacrococcygeal dorsal roots determined electrophysiologically in one rat. Note that the tail base is innervated by the first sacral root (*S1*), the mid-tail by *S3* and *S4*, and the tail tip by the first coccygeal root (*Co1*) and *Co2*, respectively

Fig. 2. Caudal plexus of one rat revealed by dissection and camera lucida tracing. Note the extensive branching of the spinal nerves. Caudal nerves going to the tail of the rat contain nerve fibers from many segmental spinal nerves

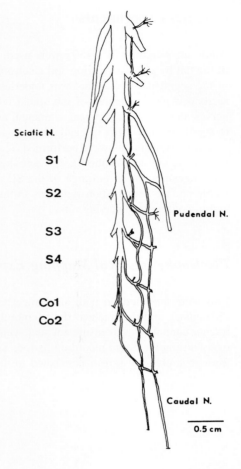

was used for supplementary doses of anesthetic. During surgery, a bolus of diluted pentobarbital was given as needed. After all surgical procedures had been completed, a continuous infusion of a diluted pentobarbital solution (10–20 mg/kg/h) was initiated at least 1 h before the actual recording began and was continued throughout the recording period. The infusion speed was adjusted to maintain a condition of areflexia with constant blood pressure and heart rate. The rectal temperature of the rat was maintained at 37.5°–37.8°C with a feedback-regulated heating blanket.

The rat was mounted on a stereotactic apparatus. A craniotomy was carried out on the right frontal and parietal bone, the dura was removed, and the surface of the cortex was covered with saline-soaked cotton balls. A vertebral clamp on the spinous process of the T1 vertebra was used to dampen the respiratory movement of the cortical surface.

Two types of mapping techniques were used: an evoked field potential technique and a unit recording technique, including single-unit and multiunit recording methods.

Evoked Potential Mapping

Field potentials evoked by electrical stimulation of the tail were used to construct topographical maps of the tail area of the SI. The stimulation electrodes consisted of three pairs of needle electrodes inserted about 5 mm deep into the tip of the tail (tail tip), in the middle portion of the tail (mid-tail), and in the place where the scales of the tail just begin to appear (tail base). The separation between the positive and negative poles was about 1 cm, with the negative pole placed rostrally.

The recording electrodes were glass micropipettes, the tip diameters of which were sharpened to 20–40 µm. These electrodes were filled with artificial cerebrospinal fluid (ACSF), which comprises NaCl 124 mmol, KCl 2 mmol, $MgCl_2$ 2 mmol, $CaCl_2$ 2 mmol, KH_2PO_4 1.25 mmol, $NaHCO_3$ 26 mmol, and glucose 11 mmol (pH 7.4) and containing 1% pontamine sky blue (K&K Laboratories, Cleveland, OH, USA). A commercial amplifier (Grass P511; AstroMed, West Warwick, RI, USA) was used to record the cortical evoked potentials (EPs) with a bandpass filter set at 3–300 Hz at a gain of 2000 or 5000. The signals were monitored with an oscilloscope and an audio monitor and were stored in a tape recorder (Neurodata DR-890). A 12-bit analog-to-digital converter card was used to digitize the cortical potential at a sampling rate of 2K online. In total, 512 peristimulus data points were collected for each stimulation cycle, and 50 cycles of cortical EPs were averaged to enhance the signal-to-noise ratio.

Stimulation parameters and recording depth were determined in a series of preliminary experiments. The recording electrode was placed in the tail area of the SI, and a constant current with a square wave pulse 2 ms in duration was used to stimulate the tail. The threshold intensities that just evoked noticeable EPs were found to be in the range of 100–200 µA, and the first component of the EP reached its maximum height at an intensity range of 300–400 µA. Subsequently, a stimulation intensity of 500 µA was used. With a repetition rate slower than once every 10 s, complex oscillatory responses were found in cortical EPs. As the interstimulus interval was shortened, the longer-latency components became smaller. The size and shape of the fast component, however, remained unchanged with a repetition rate of up to 2 Hz. Therefore, in the present study, a repetition rate of 2 Hz was chosen to emphasize the short-latency responses, which were most likely responses evoked from rapidly conducting primary afferent fibers.

A recording was made with vertical penetrations 1 mm deep in the cortex. The reasons for choosing this depth for mapping were twofold. First, this depth corresponded to the depth where most of the single-unit data were collected (see below); and second, a single large, prominent, short-latency EP was usually found at this depth following electrical stimulation of the tail.

The mapping was done sequentially with a single recording electrode in regularly spaced rectangular recording grids. The intervals between the recording points were 0.5 mm rostrocaudally and mediolaterally. Recording sequences were rostrocaudal or caudorostral rows starting either laterally or medially. At each recording point, the averaged cortical EPs were obtained for tail-base, mid-tail, and tail-tip stimulations in sequence. When a large surface blood vessel was encountered during the mapping process, this point was skipped. Data from a point farther away in the same axis was sampled instead, and the missing data point was filled in by linear interpolation. The entire process resulted in a complete set of data grids of variable dimensions, the smallest of which was 5×6, corresponding to a data collection time of 1–2 h. A representative example is shown in Fig. 3.

The stability of the cortical EPs under the conditions of the present study was tested with a control experiment. Recording electrodes of the same configuration and filled with the same filling solution were used to record cortical EPs. The electrode was placed 5 mm deep in the middle of the tail area of the SI. The exposed cortex was protected with saline-soaked cotton balls. Averaged EPs by electrical stimulation of the tail base, mid-tail, and tail tips were obtained for this point once every 0.5 h for at least 5 h. The initial latency, peak latency, and peak-to-peak amplitude of the first EP over time were compared and showed no significant changes.

The size of the tail area of the SI was estimated from isopotential maps calculated offline. An example is illustrated in Fig. 4. A 5×8 recording grid of tail stimulation-induced EP traces was obtained from sequentially recording 30 cortical points and interpolating these points. The potential changes at which a maximum EP occurred were used to construct the isopotential map. To obtain better spatial resolution, linear interpolation was used to fill in nine values between each set of adjacent data points. Two sets of gray levels were used to emphasize either the maximum amplitude of the EP or the spontaneously occurring fluctuation of the cortical potential (noise). Figure 4 shows an example of data plotted according to the maximum amplitude of the EP response. The peak amplitudes of the response in this example were 488, 408, and 288 μV, respectively, for tail-base, mid-tail, and tail-tip stimulations. These values were used as the maximum gray level, and the other levels were adjusted accordingly. A line of half-maximum amplitudes was then drawn. The area inside this line was defined as the responsive area measured with the half-maximum response amplitude. The size of the entire tail area was obtained by combining areas of the tail tip (Fig. 4c), mid-tail (Fig. 4b), and tail base (Fig. 4a) determined similarly. This is shown in Fig. 4d. Note that different cutoff levels were used for the individual maps.

Responses greater than twice the noise level were used to define the responsive area with the second method. Within each trace of the average EP, there was 20 ms of a prestimulus zone (Fig. 3). Variations of data values in this zone reflected the noise level in our preparation and the recording system. The maximal range of this variation that occurred in the prestimulus zone among all of the 30 averaged EP traces was defined as the noise level in this animal. This value was 69 μV. Therefore an isopotential line of 138 μV (i.e., twice the noise level) was drawn, and the area

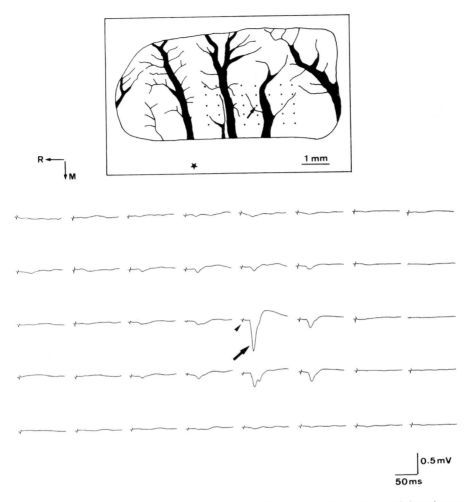

Fig. 3. Technique used for mapping the evoked field potential of the tail area of the primary somatosensory cortex (SI). **Top** Recorded points as seen from above. They comprise a 5×8 matrix in the middle of the cortical window in the contralateral frontoparietal bone. The *star* denotes the location of the bregma. **Bottom** Evoked field potential traces recorded at these points. The stimulation parameter consisted of a square wave pulse, 0.5 mA in intensity and 2 ms in width, delivered to the mid-tail of the rat at a repetition rate of 2 Hz. *Arrows* in the upper and lower panels point to the site at which the largest evoked potential was recorded in this rat. The *arrowhead* points to a stimulation artifact

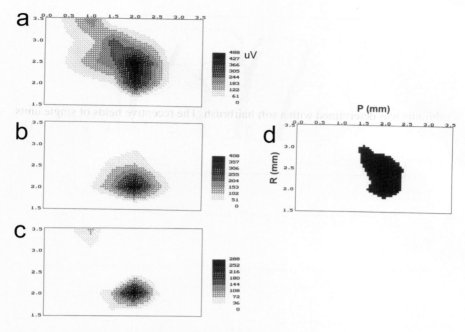

Fig. 4. Example of using the evoked cortical field potential for delineating the tail area. **a–c** Isopotential maps obtained from interpolation of cortical field potentials evoked by electrical stimulation of the tail base (**a**), mid-tail (**b**), and tail tip (**c**). **d** The tail area of this rat was determined with a combination of the three half-maximal contours

enclosed within this line was defined as the tail-base responsive area. The tail area of the rat was similarly constructed by combining the tail-base, mid-tail, and tail-tip maps.

Single-Unit and Multiunit Mapping

The preparation and recording setups used here were similar to those in the EP mapping studies. Minor modifications were made to enhance the stability. A recording chamber made of a plastic cylinder 15 mm in diameter was cemented onto the cranium with dental cement so the recorded cortical area could be covered with a layer of 4% saline agar during the recording period. A microwire sealed within a glass pipette (GL305T; MicroProbe, Clarksburg, MD, USA) was used as the recording electrode.

Multiunits could be recorded throughout the cortical depth. In contrast, stable single units were usually obtained in the deeper layers, probably corresponding to the large pyramidal cells in layer V. They were judged as single units by their all-or-none spikes.

SI Tail Area and Pain

The sequence of unit mapping was as follows. Perpendicular recording tracts began near a point 2.5 mm caudal and 2.5 mm to the right of the bregma. The response of the multiunits to cutaneous stimuli was tested at a depth of 0.9 mm. Single units were sought 0.5–1.5 mm deep in the cortex. When a clear single unit was encountered, its response to cutaneous stimuli was tested. The cutaneous stimuli used were air puffs, brushing, and light tapping. The receptive field of the multiunits was determined with a soft hairbrush. The receptive fields of single units were determined with a 10-g von Frey hair whose tip had been fire polished. The microelectrode was raised and moved 0.2–0.3 mm to another point of the cortical surface that was clear of surface vessels, and recordings began again. The sequence of recordings followed a concentric pattern surrounding the area where the receptive fields of the units covered the tail of the rat. This sequence was stopped when a full circle of nontail points had been sampled.

An example of one of the unit mapping experiments is shown in Fig. 5. The X- and Y-axes, respectively, corresponded to the anteroposterior and mediolateral

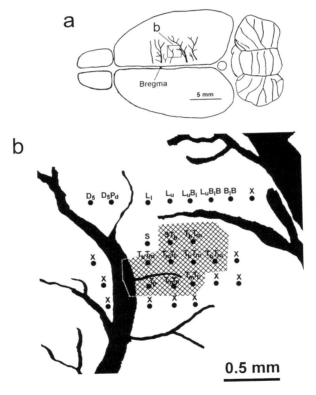

Fig. 5. Example of using the unit recording method for delineating the tail area. **a** Dorsal surface view of the brain, with the exposed vessels drawn on the right frontoparietal cortex. The *mapped area* is shown in higher magnification in **b**. Each *solid dot* represents one penetration. The tail area is *crosshatched*. B, back; D_5, fifth digit; L_l, lower leg; L_u: upper leg; P_d, paw pad; S, scrotum; T_b, tail base; T_m, mid-tail; T_p, tail tip; X, no receptive field found

Table 1. Estimated size of the time representative area of the primary somatosensory cortex of the rat determined by evoked field potentials or the unit recording method

Rat no.	Evoked field potential		Unit recording	
	2× Noise	1/2 Max.	Multiunit	Single unit
1	2.52	0.42		
2	2.08	0.82		
3	1.73	0.72		
4	0.38	0.08		
5			0.65	0.28
6			0.19	0.1
7			0.56	0.34
8			1.42	0.68
9			1.06	0.37
Average size	1.68	0.51	0.78	0.35

The estimated sizes are given as square millimeters

axes. The dots were the cortical points sampled. The receptive fields of the units are noted next to the recording points. Tail responsive single units were found in 9 of the 27 recording tracks. By interpolation, the tail representative area of this rat was determined as the cross-hatched area by the single-unit recording method.

The size of the tail area of the SI of the rat as estimated with the EP and unit recording methods are listed in Table 1.

Stereological Study to Determine the Neuronal Density of the SI Tail Representation Area

In the experiment to determine neuronal density of the SI tail representation area, two studies were performed. The first one used five male Wistar rats to localize the tail representative area of the SI with histological landmarks, and the second study used three rats to obtain the neuronal density in the tail representative area of the SI.

Location of the Tail Area in the SI

The correspondence of the tail area as recorded with multiunit mapping to the medial edge of the SI was determined by stereotaxically marking the recorded cortices. Multiunit mapping of the right mediorostral parietal cortex was performed as described in the mapping section. The recording electrode used here was the 20- to 40-µm tipped glass microelectrode filled with ACSF and pontamine sky blue. About 50–100nl of the filling solution was pressure-ejected 1mm deep into the center of the tail area by a pneumatic pump (PPM-2; Medical Systems, Great Neck, NY, USA). Within each plane, four points at equal distance occupying a square were marked stereotactically with insect pins. These marks not only helped us align the knife angle while producing the frozen sections, they were used for calculating a shrinkage factor for the tissue.

The deeply anesthetized rat was removed from the stereotactic apparatus and perfused through the ascending aorta with buffered saline followed by a fixative composed of 4% paraformaldehyde and 0.1% glutaraldehyde in phosphate buffer. The brain was postfixed overnight and equilibrated with 30% sucrose in saline. Frozen sections of the coronal plane (50 μm thick) were cut on a sliding microtome. Camera lucida drawings of the wet tissue containing the dye marks were made with a drawing tube attached to a stereomicroscope (M3Z; WILD, Gais, Switzerland). Serial sections were mounted on gelatin-coated slides and stained with thionin (Appendix A in [9]).

The SI is characterized by a thick layer IV [10]. To test the hypothesis that the tail area is located at the medialmost tip of the SI, it was necessary to find this tip. This was done by measuring the distance between the midline and the medial edge of layer IV of the left cortex in every section. The correspondence of the center of the tail area as indicated by the dye mark was compared.

The results for five rats are shown in Fig. 6. There was a close match of the electrophysiologically determined tail tip representation area labeled with pontamine sky blue (Fig. 6, arrows) with the medialmost edge of layer IV as histologi-

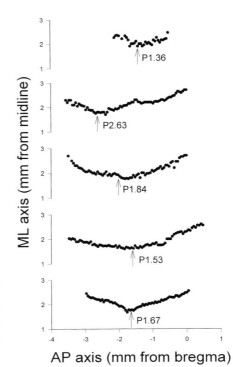

Fig. 6. Close match of the histological landmark (i.e., the medialmost edge of the layer IV boundary), with the electrophysiologically determined tail tip location (*arrows*) in five rats. The distances of the medial edge of the layer IV boundary from the midline were determined and plotted on the Y-axis for the antereoposterior (*AP*) level (X-axis) of 50-μm serially sectioned coronal sections

cally determined on the anteroposterior axis, at times with a mismatch of as small as one to two sections (<100 μm). In contrast, the points where the hind leg or trunk were mapped were both more than 500 μm away. On the other hand, the mediolateral locations of the pontamine sky blue dye marks were about 100–400 μm lateral to the edge of layer IV. Therefore, in the next experiment, we marked a 250-μm wide transitional zone here.

Stereological Determination of Neuronal Density

For optimal neuronal counting, intact, nonmapped brains were used. Rats were anesthetized with pentobarbital through an intracardial perfusion. After stereotactically marking the area with four insect pins, the brain was excised and postfixed as described above. After dehydration in an ascending ethanol series, the cortical block was embedded in paraffin. Serial 5-μm paraffin sections were cut coronally. The sections were stained with thionin [9] and mounted at 20 per slide.

A stereomicroscope equipped with a camera lucida was used to draw the contour of the sections and the position of the medialmost edge of layer IV. The sections containing the tail area of the SI of the rats was determined by the method shown in Fig. 6 and as described above. A light microscope (Vickers M41 Photoplan) equipped with a CCD camera and a video graphic printer (Sony UP-870MD) was used to observe and record the details of these histological sections. A 40× objective was used, and the final magnification as calibrated against a stage micrometer was 316× on the printed paper.

The dissector method was used to estimate the neuronal numbers [8, 11, 12]. Five equally spaced sections, each separated by 10 sections, were chosen as the dissectors. The adjacent section immediately next to these dissectors was used as the look-up section. For each pair of dissectors and the look-up section, a 750 μm wide column containing a 250-μm tail SI zone, a 250-μm transitional zone, and a 250-μm motor cortex zone (Fig. 7) from lateral to medial was photographed at 316× from the surface of the pia mater to the white matter. Prints from all five pairs of sections were reconstructed into a large montage, each about 43 cm wide and 70 cm high. The three zones and the cortical layers were marked. Neuronal nuclei were used as test objects. Only those neurons with nuclei appearing in the dissector section but not appearing in the look-up section were counted as "tops" (Fig. 8).

The area of each cortical layer (a_{dis}) in each cortical zone was measured using a digitizer. The thickness of the section (h) was calibrated to be 4.8 μm on average. The volume of each layer (V_{dis}) in the dissector section and the cortical area (A) of the sections were calculated by the following equations.

$$V_{dis} = a_{dis} \times h \quad (1)$$

and

$$A = 250 \times h \quad (2)$$

Fig. 7. Photomicrographic montage of a representative cortical section showing the use of the layer IV boundary (*arrow*) to delineate the transitional zone (*T*), the tail area of the primary somatosensory cortex (*SI*), and the motor cortex (*MC*), each of which is 250 µm in width. *Bar* 250 µm

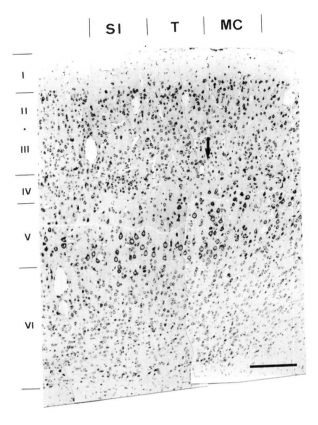

Fig. 8. Method used to determine the "tops." Vessels (*stars*) were used to align two adjacent sections. **a** Dissector section. **b** Look-up section. Those neurons that show a nucleus in the dissector section but not in the look-up section were designated "tops" (*arrowheads*). *Bar* 20 µm

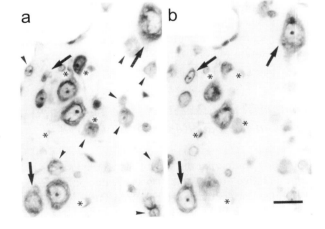

Table 2. Number of neurons in the tail representative area of the primary somatosensory cortex, transition zone, and motor cortex in three rats

Cortical layer	Primary somatosensory cortex		Transition zone		Motor cortex	
	Average	SD	Average	SD	Average	SD
I	22 429	4 946	20 159	1 962	17 351	2 669
II/III	83 083	4 038	74 467	1 047	71 892	7 070
IV	170 057	29 701	148 691	34 596	—[a]	—[a]
V	70 525	15 350	67 179	2 046	69 635	5 689
VIA	98 641	3 396	95 719	2 292	102 832	14 070
VIB	100 661	1 618	100 854	17 368	98 833	27 669
Total	84 829		78 873		75 729	

Results are the number of neurons per cubic millimeter
[a] There is no layer IV in the motor cortex

The number of neurons/mm³ (Nv) and the number per 1 mm² of cortical area (Nc) were then calculated, respectively, as follows:

$$Nv = \frac{\sum Q_i}{\sum V_{dis}} \times 10^9 \quad (3)$$

and

$$Nc = \frac{\sum Q_i}{\sum A} \times 10^6 \quad (4)$$

where Qi is the number of tops counted in each area. The numerical data were compared in each layer among the three cortical zones by repeated analysis of variance (ANOVA).

The average neuronal density was determined in three rats. The values for the volume density (Nv) are shown in Table 2, and those for the area density (Nc) are shown in Fig. 9. Our data showed that there was no significant difference among the tail region, the transitional zone, and the motor cortex. There was, on average, 84 829 neurons/mm³ and 117 750 neurons per 1 mm² of cortical area in the tail representation region of the SI.

Comparison of the Size of the Tactile and Noxious Responsive Areas in the Tail Representative Area of the SI

The study comparing the size of the tactile and noxious responsive areas in the tail representative area of the SI was conducted on six female adult Long-Evans rats weighing 250–320 g. Surgical procedures were performed under pentobarbital anesthesia (50 mg/kg IP). Ketamine hydrochloride (50 mg/kg IM) was administered

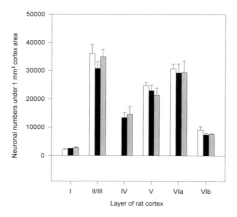

Fig. 9. Average neuronal density of the six cortical layers per $1\,mm^2$ of the cerebral cortex estimated for the primary somatosensory cortex (SI) (*back-slashed columns*), the transitional zone (*black columns*), and the motor cortex (*white columns*)

as necessary to maintain proper anesthetic depth so areflexia was maintained throughout the surgical period. The rat was mounted on a stereotactic apparatus. A midline incision was made over the skull. After retracting the skin and cleaning the soft tissue, small craniotomies were made for placement of intracortical microelectrodes. The microwire electrode arrays implanted in the right SI was a linear array consisting of 16 microwires aligned anteroposteriorly. Each microwire was a stainless steel wire insulated with Teflon (50 μm o.d.) [13]. The anteroposterior span of this electrode array was 8 mm and was centered on the hindlimb and tail regions of the SI. The electrode array was placed about 0.4–0.9 mm deep. The receptive fields of the individual channels were ascertained when the rat was still under anesthesia. A stainless steel screw (1 mm diameter) for recording the electroencephalogram (EEG) was placed on the left side of the skull at SI, P2.5, and L2.5 mm. The reference and ground electrodes were stainless steel screws located in the frontoparietal bones and over the top of the cerebellum (the mid-occipital bone), respectively. The hole in the skull and the implanted electrode were sealed and secured with dental cement. The rat was allowed to recover for at least 1 week after surgery.

Before the experiments, the rats were placed in the acrylic box that served as the recording chamber five times (for at least 2 h each day) to allow familiarization with the chamber. On the day of the experiments, the rats were anesthetized with halothane (4% in 100% oxygen) to connect the headstage of the amplifier to the electrodes and then remained in the recording chamber for 30 min to allow the anesthesia to wear off. The stimulus was generated by a CO_2 laser (medical surgical laser, Tjing Ling #2, National Taiwan University), with a 10.6-μm wavelength operating in the TEM_{00} mode (Gaussian distribution) [14]. Radiant heat pulses were applied to the skin of the mid-tail. The duration of the stimulation pulse was 15 ms. The beam diameter was 3 mm (unfocused). To minimize tissue damage, sensitization, and habituation, the stimuli were randomly applied to a local skin area of the middle part of the tail 1 cm in length. The interstimulus interval was longer than 10 s. At least 20 laser heat stimulations were made for each trial, and any jerking movements of the tail (i.e., a tail flick) were noted. The output power

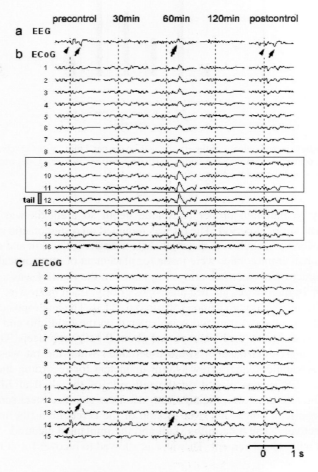

Fig. 10. Electroencephalogram (*EEG*) (*top trace*, from skull electrode) and electrocorticogram (*ECoG*) (*middle panel*, from 16-channel microwires) recorded in a chronically implanted rat. $\Delta ECoG$ (*lower panel*) is derived from the *ECoG* as described in the text. At the instant marked with the *vertical dotted line*, an 8-W, 15-ms CO_2 laser heat pulse was directed onto the mid-tail of the rat. Note the short-latency (*arrowhead*) and long-latency (*arrow*) responses [30, 31] in the evoked potential recorded in the consciously behaving rat either before (*precontrol*) or 24h after (*postcontrol*) an intraperitoneal sodium pentobarbital injection (50mg/kg). In the middle columns, evoked *EEG*, *ECoG*, and $\Delta ECoG$ recorded 30, 60, and 120min after the pentobarbital injection are shown. Note that at 60min after the injection, while the rat was still recovering from the anesthesia, the long-latency response to the noxious heat pulse returned but with reversed polarity [26, 30, 31] (*double-headed arrow*). Only channel 12 showed multiunit responses to light tactile stimulation of the mid-tail. Note the relatively larger *ECoG* responses of this channel and the channels surrounding it (*boxed channels*)

Table 3. Comparison of the size of the tactile versus noxious tail responsive area in rats chronically implanted with an anteroposteriorly oriented 16-channel linear array microwire centered on the primary somatosensory cortex

		ΔECoG, laser heat			
		Anesthetized		Conscious	
Rat no.	Multiunit, tactile—anesthetized	A[a]	C[b]	A	C
1	1	3	4	3	4
2	1	0	1	2	3
3	1	1	2	2	4
4	2	2	1	2	2
5	3	3	0	2	4
6	2	1	0	2	3
Average size	1.7	1.7	1.3	2.2	3.3*

Size is determined by the number of channels showing a response
ΔECoG, the change in the ECoG
*$P < 0.05$ vs. the anesthetized C response
[a] Short-latency primary somatosensory cortex (SI) response
[b] Long-latency SI response

was about 7–8 W. These output energies corresponded to 105–120 mJ. We examined the effect of sodium pentobarbital on the SI neural responses (50 mg/kg IP), before the pentobarbital injection (which constituted the control trial), and 30, 60, and 120 min and 24 h (postcontrol) after anesthetic administration.

The intracortical field potential (the electrocorticogram, or ECoG) and skull potentials (EEG) were recorded from the implanted microwire electrodes and screws, respectively. These neuronal activities were acquired by a multichannel neuronal acquisition processor system (MNAP; Plexon, Dallas, TX, USA). The bandpass filter used for the ECoG and EEG was 3–90 Hz. To explore a possible generator from the ECoG, the difference between adjacent channels was derived as follows. ΔECoG of channel b is:

$$\Delta ECoG(b) = 2b - (a + c) \tag{5}$$

where a and c are corresponding values of the adjacent channels.

A 16-channel microwire array was chronically implanted in six rats. Figure 10 shows a representative result of the recorded laser-heat EP and derived ΔECoG in one rat. Because the original ECoG traces of the chronically implanted electrodes might have had a common reference problem such that the evoked responses spread to all channels, we chose to use ΔECoG to estimate the size of the responsive areas. A summary is shown in Table 3. On the day of implantation, the 16-channel microwire probe was placed anteroposteriorly across a large area of the cortex that spanned anteriorly to the motor cortex, centered on the SI, and posteriorly to the occipital cortex. Individual microwires have an average interelectrode distance of 560 μm, and we found one to three tail channels (i.e., channels whose multiunit activities responded to light tapping of the mid-tail). This value did not differ significantly from the number of channels that had a short- or long-latency response to the noxious laser radiant heat applied to the mid-tail under the same anesthetic.

On the other hand, the number of responsive channels to the long-latency response was significantly greater when the rat was conscious and behaving.

Discussion

Topographically organized sensory representations exist for body parts in the cerebral cortex. Most prominent is the primary somatosensory cortex (SI). It is currently thought that inputs from the sensory epithelium during development specify and strengthen the designation of this representation [15–17]. Therefore, the relative size of the tail representation area in the SI reflects the relative number of sensory receptors compared to other body parts. On the other hand, the relative size of the SI in the total cerebral cortex is species-dependent [7, 16]. Thus, the primary cortices occupy a larger fraction of the cerebral cortex in evolutionarily simpler mammals such as rodents, whereas the association cortices dominate in advanced mammals such as primates [7, 18].

In the present study, we estimated the number of cortical neurons in the SI that are involved in processing inputs from the tail through a series of three experiments. In the first experiment, electrophysiological mapping techniques were used to delineate the size of the tail region in individual rats. The estimate ranged from 0.35 (for single-unit mapping) to $1.7\,mm^2$ (with the EP height twice the noise). Individually, the range was even wider, from barely detectable to $>2.5\,mm^2$. This is surprising in light of the general view of a stable SI. Several factors may have contributed to these variations. Methodologies may be the most important factors. Single-unit and multiunit methods detect action potential spikes, whereas the evoked field potential detects the action potential and synaptic potentials, with the synaptic potential predominating. Therefore, unit recording techniques tend to miss subthreshold signals and result in underestimates. On the other hand, all electric potential signals have a volume conduction effect because neurons are embedded and bathed in a conducting solution [19]. Therefore, the size of the source of the electrical signal may be overestimated with signals barely detectable over the noise level. Hence, a reasonable estimate should be somewhere in between. A slightly larger value than the $0.78\,mm^2$ measured with the multiunit method may be the most accurate estimate.

The second experiment of this study used the dissector method to estimate the neuronal density in this area. Our estimate showed a range of neuronal densities similar to those in the literature [20–24]. Combining the size of the area with the neuronal densities indicated that the number of cortical neuron processing sensory inputs from the tail was 94000. As stated earlier in the chapter, we first thought that the neuronal number might be relatively small, so the detailed circuitry of the tail center could easily be drawn. Describing a circuitry that involves 94000 neurons is a daunting task. Therefore, a massive simplification based on the cortical layer or cortical column, or other rules, must be used.

The primary somatosensory cortex is a key area in processing many somatosensory submodalities. Among them, pain/temperature, touch/vibration, and joint/

muscle may each be considered a larger subgroup. Supposedly, the SI is involved in the discriminative aspect of all these senses. How then should these 94 000 neurons divide their jobs, as the tail of the rat possesses all three senses? Several previous studies [25–27] addressed the representation of noxious input versus tactile input in the SI area. A general consensus is that the representation area of the two submodalities overlap and are superimposed on each other in the SI. In most of those studies, it was found that nociceptive inputs activated a relatively larger area than did tactile inputs. It is unclear, however, as to whether this reflects only that noxious stimulations are usually stronger, and stronger inputs activate larger areas than do weaker ones. In the present study, we found no significant difference between the receptive channel numbers responding to light tapping versus those responding to noxious radiant heat. It may be that the crudeness of the methodology used in the present study could not detect such subtle differences. It remains an interesting question as to how a small volume of neurons in the cortex can perform so many sensory jobs.

Although orderly topography is the most impressive initial finding, following the finding by Merzenich et al. [28] that the loss of a digit changes the topography of the digit representations in the SI, cortical plasticity of the somatosensory system has since been documented in the barrel cortex, in the nipple area, of the monkey, of the rat, and many other mammalian species studied. More recently, it was demonstrated that acute reversible deafferentation triggers immediate reorganization of the cortical representation of the whiskers [29]. Plasticity and modifiability have been thoroughly established as the normal function of the cerebral cortex. What the orderly topography represents should be considered the final result shaped by evolution and the daily life of the subject, and this topography is continually modified throughout the individual's life.

References

1. D'Amour FE, Smith DL (1941) A method for determining loss of pain sensation. J Pharmacol Exp Ther 72:74–79.
2. Hebel R, Stromberg MW (1986) Anatomy and embryology of the laboratory rat. In: Nervous system. Biomed Verlag, Worthsee, Germany, p 125.
3. Jaw FS, Yen CT, Tsao HW, et al (1991) A modified "triangular pulse" stimulator for C fiber stimulation. J Neurosci Methods 37:169–172.
4. Chapin JK, Lin CS (1984) Mapping the body representation with SI cortex of anesthetized and awake rats. J Comp Neurol 229:199–213.
5. Mitchell D, Hellon RF (1977) Neuronal and behavioral responses in rats during noxious stimulation of the tail. Proc R Soc Lond B 197:169–194.
6. Welker C (1971) Microelectrode delineation of fine grain somatotopic organization of SmI cerebral neocortex in albino rat. Brain Res 26:259–275.
7. Woolsey CN (1958) Organization of somatic sensory and motor areas of the cerebral cortex. In: Harlow HF, Woolsey CN (eds) Biological and biochemical bases of behavior. University of Wisconsin Press, Madison, WI, pp 63–82.
8. Coggeshall RE (1992) A consideration of neural counting methods. Trends Neurosci 15: 9–13.

9. Swanson LW(1992) Brain maps: structure of the rat brain. Elsevier, Amsterdam.
10. Paxinos G, Watson C (2007) The rat brain in stereotaxic coordinates (6th ed). Academic, London.
11. Harding AJ, Halliday GM, Cullen K (1994) Practical considerations for the use of the optical dissector in estimating neuronal. J Neurosci Methods 51:83–89.
12. Pakkenberg B, Gundersen HJG (1989) New stereological method for obtaining unbiased efficient estimates of total nerve cell number in human brain areas. APMIS 97:677–681.
13. Tsai ML, Yen CT (2003) A simple method for fabricating horizontal and vertical microwire arrays. J Neurosci Methods 131:107–110.
14. Yen CT, Huang CH, Fu SE (1994) Surface temperature change, cortical evoked potential and pain behavior elicited by CO_2 lasers. Chin J Physiol 37:193–199.
15. Katz LC, Shatz CJ (1996) Synaptic activity and the construction of cortical circuits. Science 274:1133–1138.
16. Kaas JH (1997) Topographic maps are fundamental to sensory processing. Brain Res Bull 44:107–112.
17. Willshaw DJ, von der Malsburg C (1976) How patterned neural connections can be set up by self-organization. Proc R Soc Lond B 194:431–445.
18. Van Essen DC (2002) Surface-based atlases of cerebellar cortex in the human, macaque, and mouse. Ann N Y Acad Sci 978:468–479.
19. Oakley B, Schafer R (1978) Experimental neurobiology, a laboratory manual. University of Michigan Press, Ann Arbor, p 121.
20. Beaulieu C, Colonnier M (1983) The number of neurons in the different laminae of the binocular and monocular regions of area 17 in the cat. J Comp Neurol 217:337–344.
21. Beaulieu C, Colonnier M (1989) Number of neurons in individual laminae of areas 3B, 4γ, and 6aα of the cat cerebral cortex: a comparison with major visual areas. J Comp Neurol 279:228–234.
22. Beaulieu C (1993) Numerical date on neocortical neurons in adult rat, with special reference to the GABA population. Brain Res 609:284–292.
23. Ren JQ, Alika Y, Heizmann CW, et al (1992) Quantitative analysis of neurons and glial cells in the rat somatosensory cortex, with special reference to GABAergic neurons and parvalbumin-containing neurons. Exp Brain Res 92:1–14.
24. Schuz A, Palm G (1989) Density of neurons and synapses in the cerebral cortex of the mouse. J Comp Neurol 286:442–455.
25. Shouenborg J, Kalliomake J, Gustavsson P, et al (1986) Field potentials evoked in the rat somatosensory cortex by impulses in cutaneous Aβ- and C-fibers. Brain Res 397:86–92.
26. Shaw FZ, Chen RF, Tsao HW, et al (1999) Comparison of touch- and laser heat-evoked cortical field potentials in conscious rats. Brain Res 824:183–196.
27. Sun JJ, Yang JW, Shyu BC (2006) Current source density analysis of laser heat-evoked intra-cortical field potentials in the primary somatosensory cortex of rats. Neuroscience 140:1321–1136.
28. Merzenich MM, Nelson RJ, Stryker MP, et al (1984) Somatosensory cortical map changes following digit amputation in adult monkeys. J Comp Neurol 224:591–605.
29. Faggin BM, Nguyen KT, Nicolelis MAL (1997) Immediate and simultaneous sensory reorganization at cortical and subcortical levels of the somatosensory system. Proc Natl Acad Sci U S A 94:9428–9433.
30. Shaw FZ, Chen RF, Yen CT (2001) Dynamic changes of touch- and laser heat-evoked field potentials of primary somatosensory cortex in awake and pentobarbital-anesthetized rats. Brain Res 911:105–115.
31. Tsai ML, Kuo CC, Sun WZ, et al (2004) Differential morphine effects on short- and long-latency laser-evoked cortical responses in the rat. Pain 110:665–674.

Key Word Index

A
Acid-evoked pain   225
Aging   100
Allostasis   184
ASIC3   225

B
β-Adrenergic   149
Blood oxygenation level-dependent (BOLD)   63
Brain   5, 184
Brain activation   100
Bruxism   184

C
Chewing   100
Cognition   37
Conditioning   167
Coping   167
Cortex   63

D
Delayed-onset muscle soreness   203
Diffuse axonal injury   26
Diffusion   5
Diffusion tensor imaging   26
Dipole model   77
Dissector method   233
DRG   225

E
Emotional memory   167
Exercise   203
eZIS   26

F
FDG-PET SPM   26
Fear   167
Functional neuroimaging   37

G
Gene expression   131

H
Hippocampus   100, 115
6-Hydroxydopamine   149
Hyperalgesia   225

I
Inverse problem   77

L
Learning and memory   100
Locus coeruleus   149
Long-term memory   131
LTD   149
LTP   149

M
Magnetic resonance imaging   5
Magnetoencephalography   77
Mastication   115
Masticatory organ   184
Mechanical hyperalgesia   203
Memory   115
Muscle afferent   225

N
Neuromodulation   37
Neuronal degeneration   115
Neuronal density   233
Neuroplasticity   37
Neuroscience   5
Nociceptors   203

P
Pain   203, 233
Prefrontal area   100
Primary somatosensory cortex   233
Protein synthesis   131

R
Rest scan   63

S
Short-term memory   131
Single-unit activity   167
Source localization   77
Spatial filter   77
Stochastic process   63
Stress   115, 184
Synaptic strength   131

T
Tail   233
Tractography   5, 26
Transcranial magnetic stimulation   37